Valuation and Volatility

Dinabandhu Bag

Valuation and Volatility

Stakeholder's Perspective

Springer

Dinabandhu Bag
School of Management
National Institute of Technology Rourkela
Rourkela, Odisha, India

ISBN 978-981-16-1137-7 ISBN 978-981-16-1135-3 (eBook)
https://doi.org/10.1007/978-981-16-1135-3

This Springer imprint is published by the registered company Springer Nature Singapore Pte Ltd.
The registered company address is: 152 Beach Road, #21-01/04 Gateway East, Singapore 189721, Singapore

Preface

Volatility is an enigma for regulators, domestic investors, FIIs (foreign investors), traders, brokers and professionals, policy makers in general. Rising volatility in markets deters governments from making public investment decisions that impact the timely generation of employment. The understanding of volatility needs attention not only due to global economic turmoil caused by biological risks or country risks, but also to build resilient investment models in modern markets. Panic behavior or reactive responses must be separated from constructive trading goals for his (her) own benefit. The arrival of high-frequency data brings newer challenges. Not all stocks would deviate the same way in markets. Trading risks are more an art than science. The right stock may become the wrong stock in 90 days from now. There are times when the valuations can go wrong not just for a surprising impact on the firms' cost of equity, but also due to unexpected growth rates. This book intends to highlight the significance of valuation of stocks which are hugely impacted by volatility to students and audiences. The manner in which equity or portfolio tests are conducted to choose time windows, sample splits, and scientific outcomes are also important to make an assessment. The book covers chapters aimed at beginners, intermediate traders who have a basic understanding of common trading terms and standard formulae. This book includes examples and business use cases on trade data.

This text is not meant to be a timeless investment. This book is not intended for advanced researchers since it deals with frameworks that are much simpler. This book is also not designed for long-term fundamentalists or short-term technicians. Instead, it offers some tools that hopefully will help to be right more often and it shows that being aware, not to panic, disciplined and make well prepared but planned moves at the desks.

Rourkela, India Dinabandhu Bag

Contents

Part I Introduction

About the Author

Dinabandhu Bag is an Associate Professor in Finance at the School of Management, the National Institute of Technology, Rourkela, and Odisha. He has taught graduate courses in financial management, portfolio management, managerial economics and business environment. He specializes in finance, cliamte and development economics. He has authored two volumes and contributed to several journals, and presented at research conferences. He has worked as the editorial board members of journals. He has worked on consulting and sponsored research projects as Principal Investigator. He has filed patents, supervised over seven research scholars for their doctoral dissertation. He has been actively involved in product development, curriculum design, laboratory development and practical modules for graduate students. He has conducted sponsored workshops for research scholars and resource person in Faculty Development Programs. He has conducted MDP (Management Development) programs for working professionals. He previously worked for over 13 years as an corporate professional across environment auditing, commercial banking, and enterprise analytics applications for banks and financial institutions. He worked with Oracle Financial Services Software Ltd, Citibank NA, GE Capital (GECIS) Ltd, and the Reserve Bank of India, and Enviro Care Ltd, respectively. He is a life member of the Econometric Society and the Institute of Bankers in India. He holds an M.Phil. in economics from IGIDR, Mumbai, MBA in banking from IIB Mumbai, and PhD in economics from the University of Mysore in India.

List of Tables

Part I

Introduction

Introduction 1

Learning objectives:

- Define and describe the interrelated aspects of valuation and volatility
- Describe major Stakeholders in the market
- Highlight the importance of volatility for all stakeholders
- Describe the classes of investors and differentiate their characteristics
- Signify the primary goals of investment and the tradeoff between them
- Distinguish the needs of individual and institutional players
- Differentiate across Horizons of investments and expected returns
- Highlight the role of Foreign Portfolio Investors (FPI)

1.1 Valuation and Volatility

Valuation plays a key role in many areas of business. Unexpected changes to cash flows induct volatility in a business. Volatility discourages business from choosing to invest in long-term assets and hold positions. As the market plunges, volatility rises. There is a lag between volatility and stock prices, because stocks may recover, but volatility levels persist. Volatility and market prices tend to exhibit an inverse relationship. The inverse relationship between volatility and liquidity of assets is predicted by market microstructure theories. It is often interesting to assume how volatility induces the valuation. For example, it may not be easy to associate the stock's volatility with the price directly, but factors exist that impact the stock's value. The simplest of all parameters being the cost of equity and cost of capital, etc. However, there are many other parameters that are important for arriving at the value of the firm, to derive the changes to firm value.

D. Bag, *Valuation and Volatility*,
https://doi.org/10.1007/978-981-16-1135-3_1

1.2 Losers

The markets are dominated by large institutional players. However, there are other players who also lose, if not directly then indirectly. The direct losers are the equity holders and the indirect losers are the lenders or guarantors, There are stakeholders or beneficiaries in the society who would benefit from the pricing of risk by the capital markets. Volatility raises the expected cost of funds that run into generating long- term fixed assets aimed at fulfilling the last mile needs of consumers, households, and the country. The magnitude of loss depends upon the volume of sales that are made in the market. Obviously, larger volumes are driven by larger losses. Timing of exit from assumed positions also matters to enumerate assume the multiplier effect of total losses.

1.3 Stakeholders

A stakeholder is an individual, a proprietary firm, a group of investors, an institution, or any other party who has interest in the stock markets. The nature of interest could be beyond monetary interest for the greater benefit of investor protection, assets of the government, properties of the society, etc. When one considers the stock market, there are a number of stakeholders. We define a stakeholder as a participant, a primary constituent who has a direct interest in the market. The stakeholder derives monetary benefits, is deeply concerned, and could be in a position to influence the prevailing conditions in the market. This will include (1) primary dealers (or market makers), (2) retail investors, (3) traders or proprietary traders, (4) foreign portfolio funds, (5) pension funds, (6) corporate players, (7) regulators or (8) sovereign governments, etc. This includes the a wide universe of stakeholders who have a significant interest in the market.

1.3.1 Investors

They are the investors and they belong to various categories (e.g. proprietary, institution, retail investors, sovereign, etc.). The restrictions on the choice of assets arise from the characteristics of individuals and entities. Individuals return needs are understood in the form of life cycle. For example, a middle-aged salaried professional does have a set of needs and risk preferences that are different from a retired widow. Few of the common constraints for the category are described here.

1.3.2 Regulators

It makes the rules for the game. It plays the role of a referee, overseer, facilitator, and licensing authority for all players. It is responsible for bringing in confidence. The hardest part for the regulator is fixing and revising numeric limits on the operating boundaries of each of these stakeholders. When there are numeric limits, it will lead to differential responses by them, and time and again, it causes spillovers, variations, and volatility.

1.3.3 Government

The capital markets continue to exist to provide an avenue for raising capital. Financial turmoil impacts the confidence of all stakeholders and is a political challenge for the government. Modern Governments aim to bring in stability and demonstrate voters' confidence by continued moral support to grasp the undercurrents of the markets. Modern governments aim to receive dividends from their own investments, disinvest, and liquidate their sovereign holdings for a premium. Governments aims to maintain their strategic sovereign interest in entities having controlling stakes. Governments are willing to bear the liquidity risks of stocks owned by them to dissuade potential buyers, viz., institutions, corporate promoters or ventures. Governments vows to protect the safety of small investors and deter dominating players from overriding others.

1.3.4 Brokers and Primary Dealers

Brokers are the major stakeholders in the capital market. Brokerage houses facilitate the trading of stocks, providing liquidity to their clients. They appoint sub brokers and dealers to improve the breadth and depth of capital markets. They are margin players who maintain their relationship with the exchanges and comply with the mandatory operational requirements of the exchange. They comply with the laws of the security regulators. They are the point of contact between investors and the market. They are the intermediaries to find a match between the needs of two or more groups of investors. Brokers maintain liquidity and bear the full risk of settlement on behalf of their clients. To continue their active status, they secure margin deposit into their account with the exchange from time to time. They continue to seek the confidence of their clients, who are the investors by themselves and must display the highest standards of professional conduct, transparency, customer orientation, discipline, and lawfulness for the timeless existence of markets.

1.4 Types of Stakeholders

The list of stakeholders was described in the previous section. The types of stakeholders are distinguished by size, reach, profit goals, or nature of interest in the markets, etc. Table 1.1 gives the types of stakeholders in the stock market.

Table 1.1 Types of stakeholders

Individuals	Retail investors & public
NRIs (PIOs)	Non-resident Indians
FIIs	Foreign investors
Corporates	Entities
Institutions	Domestic financial institutions, trusts, insurance companies
Foreign residents	Foreign individuals
Sovereign	Domestic and foreign governments
Intermediaries	Market makers
Speculators	Speculators

The common features of their investment in the market are their goals, maturity, size, riskiness, mandates, etc. Institutions do not alter their positions at short notice due to internal governance or regulatory norms. FIIs constantly monitor host country exchange rates, credit rating, and outlook and have access to advisory services. FIIs are faster to alter their positions. Few of the important features of the stakeholders are described hereafter.

1.4.1 Intermediaries

Market makers or intermediaries are capable of providing a bid price and an offer price together. The presence of market makers improves liquidity. The makers gain from the gap between the bid and offer quotes or spread. They are the option writers, makers or sellers, where the makers are an expert who takes the maximum risk and the maximum gain because he is the seller. An option seller assumes a higher level of risk, potentially facing an unlimited loss because a stock can rise. The writer or seller must provide the shares or contract in case the buyer exercises the option.

1.4.2 Governments

Government is the facilitator of capital markets. The loss to governments is in two forms, firstly, when volatility rises, it leads to lower volume of transactions at lower prices that reduce the tax revenue. A volatile market dissuades new investors to purchase new stocks and the government fails to achieve its disinvestment targets. When public assets are mispriced, it is a social cost to be borne by future generations with welfare implications.

1.4.3 Retail Investors

Small investors need to start small by entering into the market via the pre-market route. First-time small investors cannot have the expertise, experience, skill, knowledge, professionalism, size of capital, risk appetite or the holding horizon of stocks that are expected from second-time ordinary investors. Due to their lack of experience, they may not begin trading in instruments other than IPOs because if they lose, they may decide not to come back. Obviously, liquidity is an issue for them. Few markets, occasionally, provide a staggered escape route to small investors in the form of an exit route for small or illiquid stocks held. The Exchange may provide a trading window in shares of limited lots below 500 in number. Many retail investors rely on periodic income to provide for their monthly living needs, which will not be able to cope with a fall in value. The alternative is to reply on dividend-paying stocks by owning shares with high dividend yields.

1.4.4 Speculators

All day traders are not speculators. However, there They are the traders who buy and sell stocks without any intention of owning them for a long time. A speculator is a trader who comes into the stocks and moves out of stocks often for gain. Speculators are attracted towards quick profits with no real attachment to the stocks they trade. Many of the speculators may fail to achieve supernormal profits.

1.4.5 Institutional Investors

They are the funds, banks, and large or small corporates which have been described as stakeholders. Institutional investors account for a big volume of trades in Exchanges across the world. They influence the stock's movements due to their volume. They are sophisticated players who are knowledgeable. There are insurance companies, pension funds, banking systems, and nonbanking financial entities.

1.5 Goals of Investment

The investment goals are sought by investors which are of specific nature and could not be substituted across. The investment goals are more equal to their revealed preferences for combining assets. Although investors are not alike and differ in their objectives, there are few common strings attached to in the form of liquidity, tax considerations, regulatory norms, horizon, etc. Further, the goals may vary based on their religious beliefs, ethical standards, meeting social goals, etc. As the climate and circumstances change, the goals are reviewed periodically.

The risk goals are the tolerance limits on the quantum of risk they may be willing to undertake. When the expected returns are higher, the risk is higher. For individual investors, the risks averseness is the psychological mindset. However, the risk averseness depends on the time horizon and the balance between net worth and liabilities. If investors have a positive net worth, they would be prepared for more expected losses. Institutional investors are guided by their risk goals which are completely dependent on the institutional setting, the expectations of their members and the regulatory norms, etc.

Goal setting It could depend upon various factors such as the size of return, the minimum capital required, maturity, destination assets portfolio, external borrowings, number of subscribing members, and the transaction costs. A large investor could focus on branded large-cap stocks with a much larger size. A retail investor could find meaningful value in small-cap stocks, with higher expected returns and low risk and liquidity.

1.5.1 Capital Appreciation

Capital appreciation is the long-term growth and is desired in retirement plans. Long-term appreciation is realized only when investments for much longer periods exceed 5 or 10 years. The terminal goal of satisfying the future asset purchase needs of an investor is realized only when the maturities are not in the near horizon. It works on the principle of compounding. When the current earnings or capital gains from a stock are reinvested, it generates multiplicative earnings over time. This simply means that mere holding on to the stock could provide an appreciation of wealth in the future.

1.5.2 Income Earnings

The Earnings accrues from stocks which pay consistent and high dividends during the life of holding. One may think about owning blue-chip MNC stocks of large, branded corporations with a matured history and prior evidence of regular dividend payouts. The retail buyers, who are out of the workforce and retired, would choose to garner income earnings from dividends for their living expenses. The aim is to look for consistent dividend-paying stocks with good payouts.

1.5.3 Capital Preservation

Capital preservation is the need of conservative investors who are at the far end of their working life; retired who need to ensure that they do not lose capital. The safety of stocks is more critical than moderate returns. There are non profit charity organizations, Trusts, Educational institutions or government owned bodies who avail income tax exemption benefits. Such entiries aim to preserve their corpus fund. A laborer, who is out of the job or a retiree, is unlikely to recover lost capital in the market. The investor must not lose his wealth. In the market, someone's gain is someone else's loss.

1.5.4 Obligatory Investments

The requirement mandated by regulators entails corporations to invest in a variety of assets so as to maintain liquidity, preserve their net worth, and promote the development of various sectors in the economy. Many a time concessions are also granted to stakeholders when they fulfill the social sector or domestic investment norms of the home country. There exists green energy funds, agricultural sector funds, small business funds, etc. Investments into sovereign funds are made tax-free to align the funding needs of the nation by mandatory provisions to be complied by institutions. For ordinary investors, there are long-term dedicated national funds with systemic periodic contributions, the funds of which are used for national development.

1.5.5 Liquidity

The instrument is liquid when it is traded thickly in the secondary market, such as a stock or government bond. It is the easiness to find a buyer for the stock against cash. When the horizon is shorter, such as the five years, a private equity fund with a maturity of 10 years is illiquid. For retail individuals, a small part of the basket must be met through liquid sources to arrange for unseen expenses. As described in the previous section, the pension fund will aspire to experience cash outflows at a future period only when the members retire.

1.5.6 Sector Needs

Many sectors within the economy need the display of thrust and enthusiasm from capital markets. The export orientation, energy, infrastructure, real estate, etc., need specific achievement in resources toward captive growth. Capital markets are the harbingers of enhancing expanding targets of critical sectors and have to become a channel to generate resources for them. Post covid the importance of health care and pharmaceutical entities are realzied manifold.

1.5.7 Social Impact

The social goals of a citizen are met with instruments aimed at generating resources toward fulfilling the pressing needs of the society. Sustainable investment considers the promotion of environmental, social, and public welfare goals in his portfolio selection. For example, social impact investment garners capital to achieve sustainable goals. Social impact investors have the genuine intention of making a social impact alongside a reasonable return on capital. The real social impact is hard to monitor, assess explicitly. Few exchanges may report a sectoral index of sustainable investment. Few other countries have dedicated social stock exchanges. Of late there is policy emphasis on promoting non profit companies to enter into the conventional stock exchanges to raise capital. COP 26 summit in UK in 2021, brought back the renewed vigour in ESG (Environmental, Social and Governance) portfolios.

1.5.8 Climate

Investment climate is the macro policy framework and stable macroeconomic policy that reduces uncertainty and boosts investor interest. When inflation rates are high and exchange rates are volatile investing over a long-term horizon is not favorable. Excess liquidity in the banking system or when the interest rates are low could raise medium term volatility and valuations. Legal frameworks and robust recovery mechanisms via the presence of transparent markets do promote a conducive climate for fruitful investments.

1.5.9 Goals of Institutions

The institutions are driven by fixed investment policy and internal norms. They are characterized by longer horizon, periodic review of their positions at the end of the month or quarter, annual, etc. They assume bigger risks, and they do have access to borrow from external sources. They manage trading and transaction costs more efficiently. They have an understanding of hedging instruments and risk instruments. As large stakeholders in the entities, they participate in company board meetings and have superior information on the functioning of companies.

The types of institutional investors, pension plans, endowment funds, sovereign wealth funds, non-financial companies, insurance companies, etc. The institutional investors hold on the longer term when low-interest rates are lower. The matching of the stream of liabilities can cause a cash flow issue. horizon matching is important for large investors. institutional investors design return objectives, risk tolerance, tax considerations, legal, time horizon, need for capital preservation. Large-cap stocks are more popular among institutional investors.

For example, the LIC (Life Insurance Corporation of India) is a cash-rich player who has the mandate to prefer to make directed investments in government stocks of companies that are MAHARATNAs, cash rich by themselves. For example, LIC puts into oil companies, large PSUs, and also participates in money markets. It has the mandate to invest in infrastructure companies. The presence of LIC in the capital market is a harbinger that is healthy because it rejuvenates the sleeping investors. The Public Provident Fund (PPFs) serves the assets of the members of the public and the general population. The share of equities is gradually and slowly rising about 8% for PPFs in emerging markets. Overseas life insurers do deploy their funds in emerging equity markets.

1.5.10 Needs of Retail or Small Investors

Retail or small investors are public equity holders who are not recognized entities or charities, or trusts. This includes any member of the public who holds debt, equity directly or through an intermediary. They do not hold the investment on behalf of others. Retail and small investors are driven by personal ambitions, such as planning for retirement, savings for their children's education, or accumulating a reserve for larger purchases. Since they deal in a smaller size, retail investors incur higher charges on their trades, commission, and service-related fees. They assume a small proportion of the pre-market IPOs and new scrips. Their liquidity needs are seasonal during festivals. RBI laws in India permit an annual limit of $250,000 per investors in capital markets to be made by Indian citizens.

1.6 Horizons

The time horizon is the holding period to realize the gains or to comply with regulatory norms. The time horizon is important to arrive at the risk goals (Staking K and Babbel 1991). Horizon is short, medium, long or lifelong, etc. The cash inflows are expected for a shorter horizon, and for others, it may be the longer horizon, a situation of liquidity. A general insurance company could meet claims from customers any time after the policy is underwritten. A pension fund, life insurers or sovereign fund have a longer horizon.

Determining an investment horizon or term is based on the end goals of achieving a lump sum value or interim stream of earnings are needed (Oliver Wyman. 2011). The common durations are considered as short, medium, long, respectively. One may consider the period from one month to three years to be short term; three to ten years as medium term; and beyond ten years as long term.

For example, the down payment on a new house will need to use savings for at least two years. The purchase of pension for retirement is a longer time horizon. The conservative investors will choose shorter time horizon. A long-term investor may define the medium term as a holding period of one to three years. In the property market, one may regard a duration below 10 years as a medium term. Due to the unique nature of public pension plans, they are regulated largely by state and local law, federal regulation. Pension plans include contribution plans and benefit plans, respectively. Pure pension funds are different from guarantee or bonus funds. Pension funds are managed with the sole intention of serving the promised benefits to their members.

Life insurance companies invest to hedge against liabilities for two types of life policies: term policies and whole-life policy. The whole-life policy combines a severance benefit with savings. Term insurance provides death benefits. The schedule of cash outflows promised under a whole-life policy is covered in the money market. The general insurance companies, including auto, property, and casualty insurers, would need cash to settle claims which may occur in the short run. Table 1.2 describes the horizon of investment by the types of investors.

1.7 Global Portfolio Investors

Institutional investors make use of international avenues to attain the kind of a global portfolio. They believe in risk parity and purchase securities in the capital markets across geographies and markets of countries. They have requirements related to liquidity, tax, regulation, consistency with religious or ethical standards. Of course, the small and retail investors also take advantage of an international portfolio using direct equity or passive vehicles. The destination countries are ranked as very low, low, moderate, high, and very high, among others.

1.7.1 Foreign Portfolio Investors (FPI)

FPIs aim to obtain superior returns from direct investments toward host countries in emerging markets by maintaining a benchmark portfolio. A Fund, subject to its investment strategies and policies, may invest in emerging markets investments, which have exposure to the risks relating to foreign instruments. Emerging market poses many challenges. Concentrated ownership of the equity by few top promoters shareholders who are family-owned enterprises. Political, economic, and trade stability in many emerging markets (countries) have undergone reforms. Although the expected returns are higher than developed markets, the smaller size, lower

Table 1.2 Horizon of investment by the types of investors

Types of investors	Liquidity	Horizon	Compliance	Taxes
Individuals	Medium	Life cycle	None	Liable
Mutual Funds	High	Medium	Moderate	Liable
Pension Funds	Medium	Very long	Yes	Liable
Trusts	Low	Long	Negligible	None
Life Insurers	Low	Long	Yes	Liable
General Insurers	High	Short	Moderate	Liable
Banks	High	Short	Yes	Liable
Non-Banking Financial Companies	High	Long	Moderate	Liable
Corporates	Medium	Long	Moderate	Liable

liquidity, and higher volatility make it an element of diversification. Overseas funds have to comply with their internal norms, the home country, and the host country regulations, respectively.

Foreign institutional investors are allowed to invest in India's primary or secondary capital market only through the company portfolio investment scheme (PIS). fresh investment in less than 10% of the listed equity of an Indian company would be required to be carried out through the FPI route. For example, FPIs in India can purchase up to INR 1.5 Lakh crores in Debts (using Voluntary Retention Route), as per RBI limits. The FPIs in India routinely invest in direct equity, primary market equity, secondary debt, hybrids, primary market equity, etc. Further, they also invest in derivatives or ETFs in the form of index options, stock futures, stock options, and interest rate futures, etc. For example, the FIIs may not jointly or independently exceed 8% of the equity in one company stock in India. Further, the qualified foreign investors (QFI) in Indian equity markets, the aggregate holding limit of all outsiders (QFIs) in an Indian company cannot reach more than 8% of the paid-up capital of the company. FIIs do have limits on equity holding in stocks that belong to regulated sectors, e.g., banking, energy, defense, insurance, etc. These ownership limits are ensured by the government to maintain a competitive landscape.

1.7.2 Goals of Insurers

There are three classes of insurers: (1) life insurers, (2) non-life insurers, or (3) reinsurers, respectively. Life insurance companies insure the claims against death. Life insurance has a longer horizon and longer maturity products exceeding five years than non-life insurance policies, the non-life policies are held for the term of a minimum of one year. Life insurers invest in less liquid long-term assets (Winter 1991). The timing of repayments of non-life claims is more uncertain than life business. The claim rates and the claim duration of policies underwritten by life insurers determine the demand of liquidity of insurers (Zweifel and Eisen 2012). Non-life insurers maintain substantial liquidity. Reinsure institutions bear the risk of their clients who are primary insurers. Reinsurance institutions could lead to lower solvency capital requirements with higher credit risks. The host country solvency requirements are aimed at risks that can be monitored to ensure safety, liquidity and the insurer's profitability. The investment needs of reinsurers are dependent on the risk insured by them for clients.

Pension funds sell post-retirement plans to currently earning individuals. Pension plans are either defined benefit plans or contribution plans. Pension funds are long-term investors with lower needs of liquidity. The pension funds hold lower investments in equities and more investments in fixed income. Of late, pension funds have enhanced their shares to equity and enhancements of exposures to emerging markets. The factors that determine the risk goals of insurers include: their size, sources of underwriting premium incomes, horizon, etc. The goals of institutional insurers include to maintaining and preventing gaps in matching of the cash flows between invested assets and the stream of maturity and pension disbursal liabilties are known as asset and liability management (ALM). They face the challenge of meeting diversified returns that occur from classes of assets. The basic duration matching approach matches the duration of an asset with the time of liabilities. During downturns, credit insurance claims are higher. The claims of personal accidents or personal liabilities are unexpected. In life insurance and health insurance contracts, the timing of cash flows can be anticipated predicted fairly well.

The prevalence of low-interest rates can affect both the return on assets and the stream of liabilities of insurers (Albizzati and Geman 1994). The lower rates reduce the margins from low investment returns for life insurers. On the other hand, the insurers' stream of liabilities increases with low-interest rates.

1.7.3 Host Country Regulatory Issues

Regulatory requirements apply in few countries for institutional investors, such as restrictions on overseas investments, equities limits, mandated allocations to sovereign instruments, etc. In modern economies, each regulator would focus on a class of assets, when there may exist mutiple regulators for multiple asset types. Central banking compliance monitors the equity expsoure of the banking system. Pension fund regulators and insurance regulators focus on equity exposures of pension funds and insurance entities. Regulations on insurance companies are extensive. Institutional pension funds exist on the pension fund that can be invested in asset classes, equity, etc.

1.8 Summary

This chapter delved into provided a description of the basics of investments. It introduced defined volatility and discussed why volatility was important for various stakeholders and audiences, e.g., regulators, investors, market makers, society? It highlighted the classes of investors, their characteristics, investment policy, etc. It described the key challenges of types of institutional investors. It elaborated the matching principle of the horizon with return goals. It discussed the need for a global portfolio. It explained the time horizon, risk goals, and comparison between alternatives. It focused on the unique goals of individual investors as compared to larger institutions. It elaborated the matching principle of the horizon with return goals.

1.9 Questions

1.9.1 Why is there a distinction between banks and brokerage firms? Should not they be one and together for better service to clients?

1.9.2 When the number of trades is different, how do you apply the 2% rule?

1.9.3 What are the following types of funds: (1) Balanced funds, (2) ETF funds, and (3) Growth funds?

1.9.4 Among the choice of investing $50,000 in a one-year bank deposit offering an interest rate of 5% and an "Inflation-Plus" Corporate paper offering 1.5%,

a. Which one offers a higher expected return?
b. If the rate of inflation is 3% next year, which is a safer investment?

1.9.5 Give an example of the matching of the time horizon with the goals of investments?

1.9.6 Which is the lowest risk asset for below investors when the benefits are not inflation-protected?

a. investing for a three-year-old child's college admission and tuition.
b. benefit pension fund with an average duration of 10 years.

1.9.7 Which of the asset allocations is most appropriate for the pen-sion fund? Does the choice meet the objectives?

i. Return goals.
ii. Risk goals.
iii. Liquidity.

1.9.8 Briefly explain, when the household intends to purchase a house in 8 years, which of the following is the most relevant for him?

a. Time horizon.
b. Liquidity.
c. Taxes.

1.9.9 Investors may have different expectations concerning the over-all direction of the market at any time. Match the below expectations described with the appropriate strategy.

Outlook	Strategy
i. Strong Pessimistic	i. Buy stock and buy puts
ii. Strong Optimistic	ii. Buy Stock, sell call

(continued)

(continued)

Outlook	Strategy
iii. Neutral	iii. Buy put
iv. Kind of Optimistic	iv. Buy stock on margin and sell put
v. Kind of Pessimistic	v. Buy stock sell call

1.9.10 A personal investor is interested in buying Life Insurance for himself. Further, he is also interested in stocks? Can you advise him to purchase an insurance-linked fund or a fund-linked insurance?

Solutions

1.9.1 The difference between investment bank and brokerage firm is that a brokerage firm has customers who want to buy and sell things, whereas an investment bank has customers who want to uplift their money. Under Glass Stegal Act, 1933, the mixing of commercial and investment banking was considered too risky and speculative and widely considered a culprit that led to the Great Depression.

1.9.2 What are top-down or bottom-up investing styles?
Top-down investing involves looking at big picture economic factors to make investment decisions, while bottom-up investing looks at company-specific fundamentals like financials, supply and demand, and the kinds of goods and services offered by a company. While there are advantages to both methodologies, both approaches have the same goal: To identify great stocks. Here's a review of the characteristics of both methods. Bottom-up investors typically review research reports that analysts put out on a company since analysts often have intimate knowledge of the companies they cover. The idea behind this approach is that individual stocks in a sector may perform well, regardless of poor performance by the industry or macroeconomic factors.

1.9.3 What are the following types of funds, (1) Balanced funds, (2) ETF funds, (3) Growth funds?
Top-down investing involves looking at big picture economic factors to make investment decisions, while bottom-up investing looks at company-specific fundamentals like financials, supply and demand, and the kinds of goods and services offered by a company. While there are advantages to both methodologies, both approaches have the same goal: To identify great stocks. Here's a review of the characteristics of both methods. Bottom-up investors typically review research reports that analysts put out on a company since analysts often have intimate knowledge of the companies they cover. The idea behind this approach is that individual stocks in a sector may perform well, regardless of poor performance by the industry or macroeconomic factors.

(2) ETFs share characteristic features of both shares and mutual funds. They are generally traded in the stock market in the form of shares produced via creation blocks. ETF funds are listed on all major stock exchanges and can be bought and sold as per requirement during the equity trading time. Changes in the share price of an ETF depend on the costs of the underlying assets present in the pool of resources.

(3) Growth funds usually allocate a considerable portion in shares and growth sectors, suitable for investors (mostly Millennials) who have a surplus of idle money to be distributed in riskier plans (albeit with possibly high returns) or are positive about the scheme.

1.9.4 The nominal rate should equal the expected rate of inflation plus the real interest rate.

1.9.5 Let the goal is to make money for the higher education (graduation and post-graduation) of two daughters in the time horizon of 20 years from now. The Funds required are Rs 25 lakh per daughter at today's value. One will consider the expected rate of inflation during the time frame of the goal and calculating the future value of the sum as it stands today. For instance, if the average inflation is 4% per year, then for the 20 years, the projected value of future consumption becomes 2.2 times the actual value which is about 110 lakhs for two daughters.

1.9.6 Which is the lowest risk asset for below investors when the benefits are not inflation-protected?

a. While planning for investing for a three-year-old child's college admission and tuition for a horizon of 15 years from now, systemic recurring contribution plans in much secured low-risk deposit funds are advised. The alternative benefit pension funds are advised only for a much longer horizon exceeding 20 years.

1.9.7 Although it may sound impractical, all the most important are the risk goals, to begin with. However, depending on the nature and proportion of exits from the fund, liquidity could be important in the long run. Pension funds do not promise any fixed returns to their subscribers. Many a time, the returns are deep, much below the market rates in the medium horizon.

1.9.8 Briefly explain, when the household intends to purchase a house in 8 years, which of the following is the most relevant for him?
all of the following choices.

1.9.9

Outlook	Strategy
i. Strong Pessimistic	iii
ii. Strong Optimistic	iv
iii. Neutral	iii

(continued)

(continued)

Outlook	Strategy
iv. Kind of Optimistic	ii
v. Kind of Pessimistic	v

1.9.10 The total premium amount paid to the insurance company is not is invested in equity. The premium paid by the insurer goes into service expenses, life premium, and the balance in equity Investment. The amount permitted to be invested in equity by regulators is small around less than 10% of the premium. One cannot expect a higher return from insurance. Therefore, he must buy insurance for the sole purpose of life income protection.

References

Albizzati, M.-O., and H. Geman. 1994. Interest rate risk management and valuation of the surrender option in life insurance policies. *The Journal of Risk and Insurance* 61 (2): 616–637.

Oliver Wyman. 2011. *The future of long-term investing*.

Staking, K., & Babbel, D. F. 1991. *Interest rate sensitivity and the value of surplus duration mismatch in the property-liability insurance industry*. The University of Pennsylvania.

Winter, R.A. 1991. The liability insurance market. *Journal of Economic Perspectives* 5: 115–136.

Zweifel, P., and R. Eisen. 2012. *Insurance Economics*. Berlin Heidelberg: Springer-Verlag.

Volatility

2

Learning objectives:

- Begin to enumerate and differentiate across types of volatility
- Understand the Nature and Causes of Volatility
- Achieve drilling down volatility to its causes
- Highlight the importance of Liquidity factors
- Elaborate and list the alternatives of weighted to absolute volatility
- Learn and compare over numeric volatility

2.1 Volatility

Volatility is a measure of the deviation of the average returns of stock prices over time. The variance of returns is calculated from the daily returns. Standard deviation is the measure of variance from the mean of prices. Volatility shows the range in which the price varies. The range between an intraday high and an intraday low is denominated in currency terms (Andersen et al. 2003). There exist alternate notions of volatility adopted in the field.

2.1.1 Historical Volatility (HV)

Historical volatility (HV) is denominated for historical price data. It is calculated from the average deviation of the price in the sample time period. The understanding of emerging markets is stated more in terms of the trends in historical volatility. At the level of stocks, the higher historical volatility leads to higher risk tolerance or wider stop-loss levels and margin requirements. Many regulators may not permit certain categories of instruments based on their historical volatility. In general, matured markets and matured instruments would display lower historical volatility.

D. Bag, *Valuation and Volatility*,
https://doi.org/10.1007/978-981-16-1135-3_2

2.1.2 Measures

The simplest and most common measurement is the squared return of close-to-close prices. The duration-based measures such as intraday volatility or daily volatility are also used. The price range is the distance between the maximum and minimum price over a sampling interval. A synonym of the range measure includes the logarithmic range, which is the log of prices, for the difference between the highest and lowest log prices. The square of the difference between the close-to-close prices is noisy. The short interval, such as a five-minute interval of squared return is more efficient than close-to-close return, since it is more efficient, unbiased. The measures that use the open, high, low, and closing prices are more efficient than the close-to-close measure. The range-based measures are termed appropriate than the squared return of close-to-close prices. However, the range-based volatility is persistent than the squared return.

The following example uses the squared deviation of the returns from the closing prices of daily data (Table 2.1).

Table 2.1 Calculate volatility

Day	Price	Return	Deviation	Squared deviation
1	579.15	–		
2	577.95	−1.2	−1.67	2.8
3	578.6	0.65	0.18	0.0
4	580.75	2.15	1.68	2.8
5	595.15	14.4	13.93	194.1
6	580.5	−14.65	−15.12	228.6
7	570.1	−10.4	−10.87	118.1
8	577.5	7.4	6.93	48.0
9	585.4	7.9	7.43	55.2
10	593.55	8.15	7.68	59.0
11	569.15	−24.4	−24.87	618.4
12	566.6	−2.55	−3.02	9.1
13	569.15	2.55	2.08	4.3
14	585.25	16.1	15.63	244.3
15	583.7	−1.55	−2.02	4.1
16	586.3	2.6	2.13	4.5
17	589.25	2.95	2.48	6.2
18	588.6	−0.65	−1.12	1.3
19	590.1	1.5	1.03	1.1
20	588.05	−2.05	−2.52	6.3
Average		0.47	$\sum d^2_i$	1608.37
		Volatility	$\sum d^2_i/(N-2)$	89.3
		Volatility	$\sqrt{\sum d^2_i /(N-2))}$	9.45

2.1.3 Time-Varying Volatility

Time-varying volatility refers to the variation in volatility over a horizon. It is the anomaly in the nature of changes against time. Volatility is seasonal, and it displays the time effect (Copeland 1977). The volatility of stocks could be lower in the summer months compared to other months in the year. The volatility on Mondays and Friday could appear different than other days of the week. In the normal market, almost every day, the opening hours of trading may demonstrate more choppy (active) on a given day than the later part of the day when it subsides. On other days, the last hours of trading could see large fluctuations depending on specific public news or press announcements or news released to the market. It is obvious that time-varying nature of the prices is unpredictable.

2.1.4 Annualized Volatility

Volatility can be annualized to help provide this frame of reference and give some perspective. It is easier to compare volatility across a fixed 12 months horizon between stocks. To annualize the volatility, it's feasible to start calculating it over a shorter period and then, extrapolate it over the 12 months period. It is arrived from a short period interval of stock and translated to the annual period. This is also useful when the sample length of historical prices may not be available for the entire 12 months of the full year or for the entire number of trading days of 252 days. The daily volatility (σ) is multiplied by a factor that equals the root of the number of days, to arrive at the annual volatility.

For example, if $\sigma_{DAILY}= 0.01$, then the annualized volatility is $\sigma_{ANNUAL} = 0.01\sqrt{252} = 0.1587$.

The monthly volatility ($\sigma_{MONTHLY}$)can be calculated from the annual volatility by using another factor that equals the root of the inverse of the number of months.

The monthly volatility equals;

$\sigma_{MONTHLY} = 0.1587\sqrt{1/2} = 0.0458.$

2.1.5 Volume-Weighted Average Price (VWAP)

The volume of trades does not remain constant and it corresponds to the price during an interval. Volume refers to the degree of activity in the stock. The average prices could be expressed by using weights on trading volume. The volume-weighted average price (VWAP) is the ratio of the value traded during an interval to total volume traded over a particular time horizon (usually one day). It is a measure of the average price at which a stock is traded over the trading horizon. It is a benchmark that gives the comparison of the average price between stocks.

The Volume-Weighted Average Price (VWAP) is denoted as:

$$VWAP_t = \sum_i \frac{V_{it}}{V_t} P_{it} \tag{2.1}$$

where V_{it} and P_{it} are the volumes and prices at intraday time i on day t, and V_t is the daily total volume. The daily turnover over the fixed period is the denominator.

For example, the last thirty (30) days prior to the date of calculation or the last 60 trading days can be used to find the 30-day VWAP or 60-day VWAP. Many a time, the offer prices of shares are evaluated for a premium against the 6-month, 90-day, and 60-day volume-weighted average price (VWAP).

2.2 Nature of Volatility

2.2.1 Mean Reversion

The reversion refers to the retraction of stock prices back to a consensus state. When the prices move astray for some time, they may revert to their long-term average levels. Volatility is also mean reverting. The occasional periods of high volatility are pursued with spells of low volatility. It is a timing strategy to determine the range for security by computating the average price. Although it is true that in the trading room, the time duration that the stock would take to assume its long-term average levels is unknown. Further, it varies widely from stock to stock.

2.2.2 Volatility Clustering

Volatility clustering is the congregating or clustering of prices that are observed close to each other. Since volatility persists, a large change in price is followed by large changes in prices. Similarly, small changes in prices follow a small price change. It is a known time series feature of stock prices. There are extended periods of high volatility followed by a period of low volatility. High (low) volatility over the recent past is observed alongside high (low) volatility in the recent future. The markets respond to new information with large price movements. These high-volatility products may sustain for a while. The onset of volatility generates further volatility.

2.2.3 Volatility of Volatility

It is the moment of moments. It is the variance in prices given the values of one or more prices. The variance would refer to the second moment. Hence, volatility of volatility is the fourth moment. The Vol. of vol. is the standard deviation of the standard deviation of returns. When the returns are 1-year rolling or continuous, the volatility of the 1-year rolling or continuous could be normalized to an extent.

In a normal distribution, a measure of the "standard deviation of the standard deviation" is given by a relationship between sample size (n) and deviation (σ).

It is given as:

$$\sigma(\sigma)_S^2 = \sigma_S^2 \frac{N-1}{2} \times \frac{\Gamma\{(n-1)/2\}}{\Gamma(\frac{n}{2})} \tag{2.2}$$

where $\sigma(\sigma)_S^2$ is the variance of the standard deviation of a sample of N Returns, σ_S^2 is the variance of a sample of N Returns,

$$\sigma_{Si}^2 = \sum \frac{X_i - \tilde{X}_i}{N-1} \tag{2.3}$$

Γ(.) is the Gamma operator that denotes the Gamma function. An alternative explanation is the mean exceedance of the VaR (Value at Risk) limit also pointed to as the volatility of volatility.

2.2.4 Seasonality

Seasonality refers to monthly or periodic variations in levels of volatility. There is evidence that stock volatilities are dependent on days of the week. Prices could be lower on Fridays and higher on Mondays than other weekdays. Alternatively, prices could be higher on Fridays and lower on Mondays than other weekdays. When the daily return volatilities are higher on Mondays than weekdays, it refers to the week of the day effect. For example, the January month effect states that stock returns are higher in January than the rest of the year in calendar anomalies. Depending on the geographic continents, markets may be closed due to festive seasons. The annual closure and reporting of the financials of companies vary across countries. Some corporations follow the January to December, April to March, or July to June cycle. Upheavals are observed when the markets are closed down and reopen after the Easter holidays, Islamic Ramadan, or festivals.

2.2.5 Fat Tails

Returns in prices are assumed to follow the normal distribution. A standard normal distribution lies between the mean and three standard deviations (3σ) away from the mean. A fat tail is the distribution of points higher than 3σ than the normal distribution. Tail risk arises when the event occurs at more than 3σ deviations from the mean. In a standard normal distribution, the losses are explained over 99.7% within (± 3σ), where the losses are only 0.3% beyond the limit of 3σ. The normal distribution curve has a kurtosis equal to three (3.0). A distribution with kurtosis greater than three has fat tails. A heavy-tailed distribution or leptokurtic distribution covers extreme outcomes. The presence of fat tails in return distributions is ascribed to volatility clustering.

2.2.6 Trend of Volatility

The volatility of stocks increases, which is in contradiction to "mean reversal", over time, the tendency for stocks to exhibit increasing volatility over time. A strong trend upward connotes a decrease in volatility. A strong trend downward shows an increase in volatility. A reversal in trend happens when it rises. The two components of total volatility are the systemic risk and unsystematic risk; the systemic risk changes with the trend of the market. The specific and unsystematic risks are unrelated to the market. When the specific risk rises, it implies that it is more difficult to achieve risk diversification with few stocks. The volatility trend could be accompanied by a series of catastrophic events with the stock. For example, periodic disclosures of NPA (non-performing assets) numbers by Banks give jerks to specific banking stocks.

2.2.7 Long Memory

Long memory is a time series property of stock volatility. Long-memory volatility refers to a slow fall in autocorrelation between the current price and lagged price (Ding et al. 1993). This signifies the dependent nature of prices. There is correlation and dependence between successive observations that decay at a slow rate. The short -memory process exhibits a rapid decline in its autocorrelation coefficients. The short memory shows a slower or rapid fall in correlation as time increases. The long memory indicates that prices do not show immediate response to information and take longer to respond. This means historical changes in prices can be used to foresee prices. The squared returns of prices could show the presence of long memory than raw returns.

2.2.8 Co-Volatility

Co-volatility is denoted as covariance. Covariance is a statistical measure of the directional relationship between two asset prices. Covariance is of two types, Positive Covariance which implies that the asset returns have moved together in the same direction, and Negative Covariance implies that the returns have moved in opposite directions (Epps 1979). The selection of new stocks that have a negative covariance with an existing stock can reduce the overall risk. Covariance acts as a panacea against risk. The challenge is co-volatility does not remain constant, and it changes over time. Minimum variance is achieved by adding assets that have a negative covariance. Investors can select stocks that complement each other in terms of price movement.

$$\text{Cov} = \sum (R_i - \tilde{R}_i)((R_j - \tilde{R}_j)/(N-1) \tag{2.4}$$

2.3 Causes of Volatility

There are no directly attributed or straight causes of volatility. Few of the causes could be attributed to factors that impact everyone else in the market, such as liquidity, purchasing power, and tax rates. It is observed from historical trading data that volatility begins with the day beginning, and it subsides as the day moves ahead. Later, during the day, it falls flat nearing the closing hour with minor jerks due to occasional off loadings during the last hour before the end of the day closing. In between, there could be arrivals of news which may cause jumps any time during the day.

There are periodic events that occur at intervals of quarterly, half-yearly, or annually to cause shifts in expectations and price aberrations. Rare discoveries and unintended disclosures or class enforcement events may bring wider surprises and turmoil.

Policymakers make decisions which the portfolio managers cannot pre-empt. Further, the stress and impact on their assets are not pre-empted. Time on its own can generate volatility. The demand for liquidity creation is a chance to earn a premium during volatility. It is also generated when liquidity creation occurs due to the demand for some stocks that has risen, against the supply of stocks that has fallen. There is a strong relationship between the need for liquidity and the associated volatility. There exists a negative relation between the volume of stock and illiquidity or low liquidity.

2.3.1 Political Developments and Economic Indicators

The government is the most important stakeholder in the stock market. It makes policy decisions on corporate taxes, trade tariffs, indirect taxes, bi-lateral agreements, fiscal deficit, critical public safety actions, etc. The stable government makes decisions that are predictable by the industry and investors' political instability. Traders expect quarterly unemployment reports, inflation, consumer, quarterly GDP growth, and fiscal deficit. If the actual data is different than expected, the market may respond quickly.

2.3.2 Specific Risks

The specific risks impact only one stock independent of other stocks. Matters of public relations and disaster events impact one stock, business orders, and media reports. Top branded stocks are prone to public relations and media (PR) fiasco. Company news of strong earnings reports or new product launches makes the stock attractive. The negative news about the stock data breaches or allegations against top management product recall and physical damages to life and property will impact the stock price.

2.3.3 Leverage and Volatility

Leverage enhances the size of the investment supported by borrowed capital. Leverage and margin are not exactly the same. Pure leverage refers to the pure borrowing of cash. Under a margin account, the leverage is borrowing from a broker at a fixed interest rate to purchase securities in anticipation of gains. When the mandated requirements of margin are higher, it reduces the volatility of the market. The lower margin requirements tend to increase volatility. Leveraged trading can magnify both gains *and* losses. In margin trading, the losses get magnified. If the price moves against the investor, which is the multiplier effect, his or her loss is much greater than when there is no leverage. When the leverage is five times, a 10% fall in prices can cause a 50% erosion of the margin money.

2.3.4 Liquidity

The market-wide phase of liquidity shortages is observed across the span of institutional investors, foreign investors, and corporates. The need for liquidity can arise due to varying exchange rates, repo rates, or during cycles of advanced tax payments. All stocks do not convey the same degree of liquidity. The characteristics of stocks distinguish them from being less liquid than many other liquid stocks. For example, stock splits, share buybacks, mergers, or regulatory actions against companies are instances that trigger a liquidity crisis. Political turmoil also causes a liquidity crisis.

Small stocks display lower liquidity than larger stocks. The demand for liquidity can cause excess volatility in a stock. Liquidity is counted as the number of days for the buyer to be able to escape and exit by sell off. The bid-ask spreads are too low reasonable to exit and rather wait longer. The IPOs are listed and traded on the floor, and many investors aim at gains on the first day of the first minute of the listing. The first-day first-hour exit is driven by the need to maintain liquidity. Therefore, the takers on D day are the ones who have horizons that are different from IPO subscribers. First-time retail beginners have limits on their IPO holding and may decide to exit on the Day. There are no physical penalties on big institutional investors for exceeding such limits. FIIs make two kinds of given routine and serious reviews of their portfolio.

2.3.5 Integrated Markets

Modern economies have undergone perfect integration of the financial markets across the globe. Today's markets are globally connected including equity markets. Capital flows across currencies and stocks or bonds. For example, NASDAQ, LSE, and Nikkei display simultaneous influence on BSE in India. The degree of daily influence between Nikkei and BSE could be 1%. The daily impact is observed more within the geographic continent or within Asia. For example, for the global investors to hold cross-country stocks, they acquire host country currencies to purchase the stock in the denominated currencies of host country companies. Crude oil price and dollar rate cause upheavals in many markets simultaneously.

RBI laws in India permit an annual limit of $250,000 per investors in capital markets to be made by Indian citizens. Major events such as trade wars, political upheaval, and regime changes could cause changes to stock volumes, currency, or investments between countries. There is hardly a day that passes without at least one event related to the globe. The market volatility is managed with globally diversified portfolios.

2.4 Volatility Measurement

Volatility is not measured in real time. It is an assessment of the returns calculated over a historical sample period. A longer sample window could reduce the sampling error in assessed value. The accuracy becomes more acute as the sampling frequency increases. A good model can summarize the measurement less influenced by the sampling frequency, sampling interval, or sampling errors. The mean reverting behavior of volatility has been questioned by proponents of antithesis. Past volatility predicts future volatility. A lower sampling period can improve the accuracy of measurement. For example, if a stock has actual volatility of 30% on 15-min interval prices, a 0.1% bid-ask spread can alter the accuracy by 2%. The class of models includes stochastic models or latent volatility models. The arrived values are conditional or absolute determinants of the underlying nature.

2.5 Summary

This chapter provided standard and feasible definitions of volatility. This chapter gave the formula for the measurement of types of volatility. It explained the time-varying nature of volatility. This described historical volatility. Specifically, it elaborated time-varying volatility. It highlighted calculating reported measures of annualized, annual, and weighted volatility. It threw light on co-volatility and covariance. It described weighted volatility and fat tails. It highlighted the positive and negative nature of volatility with liquidity, leverage, etc. It explained the causes of volatility.

2.6 Questions

2.6.1 When a stock has returned 1% in 6 months, calculate the annualized return.

2.6.2 Calculate the annualized volatility for ABC Stock from the following data. The ABC stock has experienced daily price changes 1%, 2%, 3%, 4%, and 5%.

2.6.3 Calculate the annualized volatility of ABC stock.

$\sigma_{\text{annualized}} = \sigma_{\text{daily}} \times 252^{1/2}$.

$\sigma_{\text{annual}} = \sigma_{\text{weekly}} \times 52^{1/2}$.

2.6.4 Calculate Covariance of the Portfolio from the Data Given in Table 2.2 for Two Stocks.

Table 2.2 Price of two stocks

Date	$PRICE_1$	$PRICE_2$
1-Sep-19	579.15	83.65
2-Sep-19	577.95	84.45
6-Sep-19	578.6	84.85
7-Sep-19	580.75	83.7
8-Sep-19	595.15	84.5
9-Sep-19	580.5	83.6
12-Sep-19	570.1	82.8
14-Sep-19	577.5	83.75
15-Sep-19	585.4	84.5
16-Sep-19	593.55	85.15
29-Nov-19	569.15	79.25
30-Nov-19	566.6	77.45
19-Dec-19	569.15	76.4
27-Feb-19	585.25	114.55
28-Feb-19	583.7	115.85
1-Mar-19	586.3	113.25
2-Mar-19	589.25	109.6
3-Mar-19	588.6	110.55
6-Mar-19	590.1	109.75
7-Mar-19	588.05	108.6

Solutions

2.6.1 The 12 months Return in a year and Divide by the 6 months, the Annualized Return of 2.01%.

Annualized $\sigma_{annual} = 1.0 \times (12/6)^{1/2} = 2.01\%$.

2.6.2 The standard deviations of the sample returns are σ_{daily}= 1.58% and $\sigma_{annualized} = 252^{1/2} = 25.08\%$.

The annualized volatility for ABC Stock is 25.08%.

2.6.3 $\sigma_{annualized} = \sigma_{daily} \times 2521/^{2}$.

2.6.4 $\sigma_{annual} = \sigma_{weekly} \times 52^{1/2}$.

When the daily volatility, σ_{daily}, of a stock is 1.2%. Multiply this by the square root of 252, $\sigma_{annual} = 1.2 \times 252^{1/2} = 19.05\%$ (Table 2.3).

Table 2.3 Covariance of two stocks

Date	$Price_1$	R_{1t}	$Price_2$	R_{2t}	Variance(R_{1t})	Variance(R_{1t})	Covariance
1-Sep-19	579.15		83.65				
2-Sep-19	577.95	−0.21%	84.45	0.96%	0.00	0.03	−3.261E-05
6-Sep-19	578.6	0.11%	84.85	0.47%	0.00	0.02	4.856E-05
7-Sep-19	580.75	0.37%	83.7	−1.36%	0.00	0.01	2.495E-05
8-Sep-19	595.15	2.48%	84.5	0.96%	0.03	0.03	0.0007327
9-Sep-19	580.5	−2.46%	83.6	−1.07%	−0.02	0.01	−0.0001961
12-Sep-19	570.1	−1.79%	82.8	−0.96%	−0.02	0.01	−0.000159
14-Sep-19	577.5	1.30%	83.75	1.15%	0.01	0.03	0.0004228
15-Sep-19	585.4	1.37%	84.5	0.90%	0.01	0.03	0.0004073
16-Sep-19	593.55	1.39%	85.15	0.77%	0.01	0.03	0.0003953
29-Nov-19	569.15	−4.11%	79.25	−6.93%	−0.04	−0.05	0.0020236
30-Nov-19	566.6	−0.45%	77.45	−2.27%	0.00	0.00	1.345E-05
19-Dec-19	569.15	0.45%	76.4	−1.36%	0.01	0.01	2.915E-05
27-Feb-19	585.25	2.83%	114.55	49.93%	0.03	0.52	0.0151415
28-Feb-19	583.7	−0.26%	115.85	1.13%	0.00	0.03	−5.21E-05
1-Mar-19	586.3	0.45%	113.25	−2.24%	0.01	0.00	−1.892E-05
2-Mar-19	589.25	0.50%	109.6	−3.22%	0.01	−0.01	−7.927E-05
3-Mar-19	588.6	−0.11%	110.55	0.87%	0.00	0.03	−4.845E-06
6-Mar-19	590.1	0.25%	109.75	−0.72%	0.00	0.01	4.064E-05
7-Mar-19	588.05	−0.35%	108.6	−1.05%	0.00	0.01	−2.152E-05
Average		−0.09%		−1.89%		Covariance	0.001

References

Andersen, T.G., T. Bollerslev, F. Diebold, and P. Labys. 2003. Modeling and forecasting realized volatility. *Econometrica* 71: 579–625.

Copeland, T.E. 1977. A probability model of asset trading. *Journal of Financial and Quantitative Analysis* 12 (4): 563–578.

Ding, Z., C.W.J. Granger, and R.F. Engle. 1993. A long memory property of stock market returns and a new model. *Journal of Empirical Finance* 1 (1): 83–106.

Epps, T. 1979. Co-movements in stock prices in the very short run. *Journal of the American Statistical Association* 74: 281–298.

Realized Volatility

3

Learning objectives:

- Define, explain, and calculate realized volatility
- Gather the empirical forms of Realized Volatility
- Learn related RV concepts and Co-volatility
- Exemplify and elaborate the techniques of estimation
- Figure out the use(s) of RV relevant strategy

3.1 RV

Realized volatility (RV) is the cumulatively summed squared returns drawn over a consecutive window of small and fixed time intervals (Admati et al. 1988). It is arrived as the summation over available intraday squared returns. Realized volatility is determined ex-post from the deviation in returns (Bandi et al. 2008). The mean return is an a-priori expectation around the observed price in anticipation of an average, whereas realized variance is derived from calculated historical returns. The squaring operator turns negative returns into a positive number, which is RV.

3.2 RV Calculation

RV is the daily (i.e., open-to-close) realized volatility.

$$R_t = log\left(\frac{P_n}{P_{n-1}}\right) \tag{3.1}$$

D. Bag, *Valuation and Volatility*,
https://doi.org/10.1007/978-981-16-1135-3_3

where

R_t = realized return,

n = a counter representing each trading day,

P_t = Closing price on trading day t, and .

P_{t-1} = Closing price on the previous trading day t-1.

The alternatives are holding the stock for 1 day, 2 days, 1 week, 1 month, etc.

$$RV = \log\left(\frac{P_n}{P_{n-1}}\right)2 \tag{3.2}$$

where,

RV_t = realized return,

where n varies from 1 to T.

For n = 1, $RV_1 = \log(\frac{P_1}{P_0})^2$

For n = 2, $RV_2 = \log(\frac{P_2}{P_0})^2$

For n = 20, $RV_{20} = \log(\frac{P_{20}}{P_0})^2$

Therefore, RV_n is independent of RV_{n-1}.

While calculating intraday variance, in a given day, log prices are sampled at equidistant intervals from intraday prices ($P_{t, n}$).

Here, t refers to time measured in minutes, and P_t refers to the log closing price of day t,

where t = 1, …, m, with m the number of increment intervals.

The increment returns are.

$$r_{t,n} = \log\left(\frac{P_{t,n}}{P_{t,n-1}}\right) \tag{3.3}$$

From these intraday returns, the realized variance is given as

$$RV_t = \sum_0^n r_{t,n}^2 \tag{3.4}$$

RV_t is the realized variance at time t, and $r_{t,n}$ is the realized return at interval n.

Table 3.1 demonstrates an example of calculating RV.

The annualized RV is derived from the daily RV by multiplying it with the number of trading days/weeks/months in a year. The expectation of returns is independent of the frequency of sampling and depends on the window of the price

Table 3.1 Calculation of RV

Day 1			Day 1			Day 1		Day 1	
Date Time	Price	Log (Price)	$r^2_{t,n}$	Price	RV	Price	RV	Price	RV
1-Apr-13–9:07:41	1	0.0				1		3	
1-Apr-13–9:17:07	2	0.3	0.09	2		5	0.49	6	0.09
1-Apr-13–9:34:17	4	0.6	0.09	4	0.09	9	0.07	8	0.02
1-Apr-13–9:49:16	6	0.8	0.03	6	0.03	11	0.01	11	0.02
	Daily RV		0.21		0.12		0.66		0.13

data. A simple mean is independent of the number of data points but it depends upon the length of time. Practitioners distinguish between volatility: *actually* often referred to as historical or realized and *implied*, which is derived from the prices of options. Implied volatility is not as much of interest to all unless the investors have exposed themselves to derivatives. Actual return is realized only when one sells off securities held every now and then. Actual return is not a symbolic return or implied return. RV is positive since the square of any negative number is positive, when returns could be negative. If there are no trades between two intervals, there are any movements, then the returns and volatility are zero. If the stock price remains stable, the volatility is zero. Hence, RV gives a closer picture of the underlying movements in price.

3.3 Benefits of RV

The practical benefits of RV include the following:

1. RV aligns the weights of allocation across assets based on conditional expected return.
2. RV is squared returns. RV gives positive numbers.
3. It gives the comparison of assets when actual returns are higher; the calculated RV is also higher.
4. It is useful in determining the conditional return variance from past values of realized volatilities.
5. The variants of RV could significantly reduce or eliminate microstructure noise.

3.4 Properties

The finite sample properties and its large sample (asymptotic) properties of RV documented in previous studies (Andersen 2003; Hasbrouck 2003) include the following:

1. RV mimics daily integrated variance that is noise dependent.
2. RV is a consistent estimator of the true, integrated variance or intra-daily squared returns.
3. RV follows a continuous, correlated distribution process.
4. The logarithmic RV is a persistent, stationary process.
5. It is not a consistent estimator of pure variance.
6. RV is sensitive to extreme values, frequency, and span altogether.
7. RV is not a remedy against pure microstructure noise.

3.5 Types of Realized Volatility

Due to its nature, RV has seen attempts of implementation in modern high-frequency data of both forex and equity markets. The types of RV differ with respect to sampling intervals. Apart from the types with respect to sampling intervals (medium or small), the manner of its implementation also varies. The comparing of the RV of stock with any benchmark can show the stability of a stock. When RV is lower, it indicates a more predictable nature. The following forms of field implementation of Realized volatility have been discussed.

3.5.1 5-min Intraday Realized Volatility

The 5 min interval intraday realized volatility is considered a true measure of historical volatility (Andersen et al. 2003). The RV is the aggregated squared returns over small intervals of 5 min. This approach uses a realized volatility series constructed from a 5-min log of prices. The 5-min RV remains positive and increasing and easy to compare the traverse.

3.5.2 Median-Based Volatility

The median-based volatility is an alternate measure of historical volatility introduced in Andersen et al. (2003). The median measure is not influenced by extreme values and skewed distribution or asymmetrical distribution.

3.5.3 Range-Based Volatility

The range-based volatility represents a measure suggested by Parkinson (1980). It takes into account daily high and low prices which give information more than just the closing prices.

$$\sigma_t^2 = \log\left(\frac{H_t}{L_t}\right)^2 \tag{3.5}$$

where $\sigma_t^2$is the range variance for day t, and H_t is the daily high price, and L_t is the daily low prices, respectively.

3.5.4 Volume–Weighted Volatility VWV

VWV is a realized measure of the daily demand-based volatility. The VWV is the standard deviation of volume-weighted average price (VWAP), weighted by the proportion of the volume of shares. The VWAP is defined in Chap. 2. The volume-weighted average price is the division of the traded value to the total traded volume, over one day. The VWV is represented as a normal distribution. Weighting prices with trading volume makes VWV efficient because volume possesses more information than prices. It is given as:

$$\sigma_{VWV}^2 = \sum_i \frac{V_{IT}}{V_T}(p_t - VWAP_t)^2 \tag{3.6}$$

where $VWAP_t = \sum_i \frac{V_{it}}{V_t} P_{it}$

where σ_{VWV}^2 at time i on day t.

VWAP is the volume-weighted price, V_{it} and Pit are the volumes and prices at intraday time i on day t, and.

V_t is the daily total volume.

The quality of the information in VWV calculation is superior. It is a form of realized volatility indicator.

VWV is less volatile than the other measures (Ko et al. 1995). VWV is more appropriate for statistical tests since it does not follow fat tail distribution like ordinary volatility. Volume can be misread because large volumes throw up big tops and bottoms.

3.5.5 Realized Co-Volatility of a Portfolio

Realized covariance is more general to RV because the covariance of returns between stocks is the correlations that exist among the leads (R_{t+1}) and lags (R_{t-1}) of Returns. The difference between synchronous (simultaneous) co-volatility and asynchronous co-volatility is significant. Asynchronous correlations across stock prices are achieved by correlating with lagged prices (Stoll, 1990). The one-factor (CAPM) pricing model can explain synchronous correlations. At a portfolio level, the RCOV includes covariance computed ρ_{ij} for stocks i and j, existing within the portfolio.

For two given assets, lead–lag estimator for RCOV with m lags is given by

$$RCOV_t = \sum_{t-m}^{t} r_{t,i} r_{t-m,i}/N - 1 \tag{3.7}$$

where $r_{i,t}$ and $r_{j,t}$ are the intraday standardized returns for asset i and j at time t and lag m, respectively, and.

N is the number of traded days.

3.5.6 IPO R-Volatility

The IPOs, which are the Initial Public Offerings, eliminate concentrated shareholding of individual owners of a company by distributing ownership to members of the public at large. IPOs are oversubscribed many times which is due to large investor participation. Table 3.2 gives the categorization of IPO stocks to classify into sizes based on their issue size. There is a consistent increase in volatility of Very Large IPOs, which is observed specifically, in the post-period sample. The post-periods sample is different from the pre-period sample and the periods refer to the second mid-year and the first mid-year.

The large participation of institutional investors with shorter holding period and immediate liquidity needs may also pose volatility risks. The undervalued IPOs may draw a mad rush into their subscription and liquidation on the day of listing.

3.5.7 Hypotheses

Properties of RV are derived from the results of tests applied to real stock trade prices. One may come up with more number of hypotheses than described here. The long-range dependence behavior of RV has not been observed in empirical results. Early studies (Carvalho et al. 2006) did not find any evidence of long memory in realized volatilities in the Brazil stock exchange. The intraday persistence in RV is of interest to readers.

Table 3.2 Classification of stocks by issue size

Issue_Size $\leq$ 500 MM	Very small
500 MM < Issue Size $\leq$ 1BB	Small
1BB Million < Issue_Size $\leq$ 2BB	Medium
2BB < Issue_Size $\leq$ 5BB MM	Big
Issue_Size > 5BB	Very large

Note IPO volatilities are larger for large issue sizes
MM—Million, BB—Billion
Source Author's classification using SEBI Listed data

Hypothesis 1: RV is stationary (Log RV is Gaussian Normal).

Autocorrelation is the tendency for recent past prices that influence future prices.

Hypothesis 2: RV is causal or bi-directional.

Causation tests are the test of a bi-directional relationship between two variables. In this case, RV may influence expected returns, or expected returns may influence RV.

Hypothesis 3: Persistence of Volatility is significant.

RV follows univariate processes.

3.5.8 Sample Length

The window for an optimal sample length is determined from the feasible window of implementing a RV calculator in the field. A longer sample window has the benefits of accuracy. Small interval sampling can minimize the errors due to sampling because it controls for the non-synchronous trading effects since there is no simultaneity bias. The sampling frequency may range from 30 and 60 min since it is practical to arrive at from high-frequency data. Similarly, 15 min intervals have also been used. Although it is much lower than 5-min frequency suggested in the literature, the trading desks will need to have a strategy to take buy/sell actions on a real-time basis.

The number of sample observations needs to be at least 10 times the number of stocks in a portfolio (Madhavan and Panchapagesan, 2000). Few tests have suggested calculating RCOV with a sparse sampling frequency of 15 or 20 min onto sampling across equally spaced 5 min intervals. The long-memory influence reduces with the increase in the sampling interval (lower intervals). However, shorter intervals can yield erratic patterns of unpredictable volatility measurement. The number of records in the sample will depend upon the ratio between the 5-min RV and the 15-min RV. An average of the RV over a longer time window reduces the influence of sampling error. The average RV spikes up and subsides beyond 140 min during the day. The number of data points recommended to calculate RV is higher than 20 observations.

3.6 Empirical Model

RV tests are completely independent of CAPM one-factor micro structure pricing tests. RV is incapable of testing the significance of the bias in efficient markets. The one-factor model (CAPM Model) is synchronous with the market portfolio. RV is negatively correlated with return and positively correlated with volumes and volume imbalances (Chan 1992). For example, to investigate the absence of synchronous trading effect or simultaneity bias on the daily volatility, the following simple model gives rise to.

$$RVOL_t = \alpha + \beta RVOL_m + \varepsilon \tag{3.8}$$

where RVOL is the realized variance for the day. The parameters (β) and degree of fitting (R^2) indicate what proportion of the intraday RV can be explained by the RV generated over the first m intervals. For example, if the market closes at 3.30 PM on a given day, the volatility at 15-min interval before, i.e., 3.15PM is the explanatory variable. The test of this model is undertaken by introducing small interval lagged values for RVOL to detect the degree of persistence.

Realized volatility could occur to correlation in stocks prices due to the arrival of good and bad news or volume imbalances. Volume plays a role in driving RV. According to the dynamic return–volume relation of an individual stock, one specifies volatility model as;

$$RVOL_{i,t} = \alpha + \beta_m RVOL_{t-m,i} + \gamma_i ORDIMB_i + \delta_i VOLUME_i + \varepsilon \tag{3.9}$$

where $RVOL_{i,t}$is the intraday realized volatility in normal trading, $OrderImbal_{it,}$ is the pre-open order imbalance of stock i, and $VOLUME_{i,t}$ is the pre-open traded volume of stock i.

Further enhancement of Eq. 3.4 to a news event hypothesis formulation will lead to:

$$RVOL_{i,t} = \alpha + \beta_m RVOL_{t-m,i} + \gamma_i ORDIMB_i + \delta_i VOLUME_i + \delta_{it} EVENT_{k,t} + \varepsilon \tag{3.10}$$

where θ is the parameter for the Event k at time t.

The Granger causality tests for Log $RV_t$are also presented in Table 3.4 which infers the absence of bi-directional causality. The Granger hypothesis is rejected and there exists no bi-directional causality. The autocorrelation and stationarity hypothesis is also rejected. The RV is non-stationarity in nature.

Table 3.4 provides the estimates of Realized Volatility (RVOL) using lagged indicators in two versions of 1. POOLED regression and 2. SUR (Seemingly Unrelated), separately. The model results find Pre-Opening Order Imbalance Ratio and Pre-open Trading Volume which is positive and significant.

Table 3.3 Results of Granger-Causality tests Log RV_t

Hypothesis	DW	F-Stat	R2
F-Ratio			
RV-Return	2.118	5.573(1, 120)	0.119
6.215(3, 120)			
Return-RV	1.912	0.007(1, 120)	0.003
0.1612(3, 120)			

Note the p-values of the Granger causality tests. The null hypothesis is accepted at the 99% level

Table 3.4 15 min realized volatility with pre-opening

Dependent variable	Sample1 (six months)	Log RVt	Sample 2 (One Year)	
Estimation Method	1. Pooled	2.SUR	1.Pooled	2. SUR
Intercept	0.001 (0.6)	0.001 (0.6)	0.01(0.2)	0.01(0.2)
RVOL$^+$(23)	1.806 (<0.0001)	1.81 (<0.0001)	0.7 (<0.0001)	0.7 (<0.0001)
RVOL$^+$(22)	0.388 (<0.0001)	0.388 (<0.0001)	0.4 (<0.0001)	0.4 (<0.0001)
Order imbalance	–	–	8.61E 06 (<0.0001)	8.61E 06 (<0.0001)
Trade quantity	4.1E 06 (<0.0001)	4.1E 06 (<0.0001)	3.5E 06 (<0.0001) 3.5E 06 (<0.0001)	
NT (Total observation)	1580	1580	3381	3381
N (tickers)	104	104	104	104
(Trading days) T	14	14	30	30
SSE	23.05	23.05	382.6	382.6
MSE	0.01	0.01	0.1	0.1
R^2	0.92	0.92	0.7	0.7

1. + Number in parenthesis indicates the lag number of the indicator. (21) Indicates 21times of 15 min which is the RVOL at 1.15AM. (22) Indicates 22 times of 15 min each which is the RVOL at 10.30AM. 2, Order Imbalance is significant and positive, which implies an increase in volatility due to imbalance

3.7 Using RV

The RV is used to arrive at a longer-term price risk for 1 month or more longer. The RV numbers alone do not provide the direction of upward and downward trends in price movements. However, RV plays a role for buy/sell decisions in trading. For example, one can choose an RV threshold limit that achieves a given level of Value-at-Risk (VaR) target. One can compare the target RV of a stock to decide on allocation.

There are few options in the hands of a trader who observes RV and intends to perform a strategy. For instance, using historical calculated RV, (i) he could escape the onset of volatility by resale before it falls back on him, (ii) he can fix stop-loss limits for resale or fix limit order price to buy or sell, lastly, (iii) he could observe the 90th percentile RV limit to resale or the 70th percentile RV limit to buy, or (iv) he could revise the weights of allocation to his portfolio in the short run, etc.

If an investor finds the extreme price to be higher than the 90th percentile value, even when the actual number of losing trades is small, he has to underplay a mixed strategy. One way to increase the odds in the investor's favor is to sell at a level above the 90th percentile. He may decide to buy at a price level below the 90th percentile and sell below the 90th percentile, making a gain of 20% straightforward. Although it appears so, RV is not a day trading strategy.

3.8 Summary

This chapter detailed a review of the main features of RV. It explained the derivation and practical applications of RV with examples. It exemplified the calculation of RV. A simple empirical test model was presented. Relevant hypotheses were described. The sampling issues and critical properties of the RV estimator were highlighted. Issues relating to the nature of RV, tests of persistence, or stationarity behavior of realized variance were also described. It demonstrated the limitations of RV. It highlighted RV strategy.

3.9 Questions

3.9.1 How to calculate RV?
3.9.2 Does RV depend on the holding period?
3.9.3 What is an average RV? What is Short interval RV?
3.9.4 Is there any relation between Return and RV?
3.9.5 Will RV explain the mean–covariance asset allocation principle?
3.9.6 What is microstructure noise?
3.9.7 Is RV sensitive to the number of data points or frequency of sampling?
3.9.8 Is RV sensitive to *extreme* values in the sample?
3.9.9 Why is IPO volatility higher than normal?
3.9.10 Why is large-cap stock volatility different from small-cap volatility?
3.9.11 Why is RV more accurate than a simple static mean for short interval high-frequency data?

Solutions

3.9.1 The formulae for RV are given as $RV = \log(\frac{P_n}{P_{n-1}})^2$
3.9.2 RV does not depend upon the period of holding. RV is sensitive to the frequency of sampling and interval over which it is measured.
3.9.3 The average RV is the average over a known short interval for days of the observation window.
The short interval RV is the squared realized return over an interval of 1 minute or 5 minutes.

3.9.4 RV is negatively correlated with the ex-ante return.

3.9.5 A risk-adverse investor follows mean-variance optimization strategy.

3.9.6 Market microstructure hypothesis believes that the risk appetite of an individual investor varies with his/her own utility derived from return for a given risk. Information arrives at different frequencies for different assets; therefore, microstructure effects that are related to the non-synchronicity in price formation should be incorporated. Asynchronous trading could introduce a bias in RV estimates.

3.9.7 Yes, RV is sensitive to the frequency of sampling and also total length of the sample period.

3.9.8 Yes, Extreme values give jerks to RV. However, RV is as it is, and outliers are included in the calculation.

3.9.9 There is no direct answer. IPOs are oversubscribed many times which is due to large investor participation. Similarly, the prevalence of undervalued IPOs may draw a mad rush into their subscription and liquidation on the day of listing. Institutional traders exit on the D day of listing.

3.9.10 Large caps invite more attention from the market players. The large participation of institutional investors with shorter holding periods and immediate liquidity needs may also pose volatility risks.

3.9.11 Volatility is due to the deviation between two price points from the mean. The deviation from mean gives rise to a smaller or larger number that depends upon the calculated mean of prices over a period of a few time intervals. Such deviation over a fixed period mean could be higher or lower depending upon the value of the last calculated mean. Volatility in returns is the gain realized by an imaginary investor. Therefore, a short interval gain between two consecutive time points is more realistic than the distance from the tiptoeing mean for high-frequency trades.

References

Admati, A.R., and P. Pfleiderer. 1988. A theory of intraday patterns: Volume and price variability. *Review of Financial Studies* 1 (1): 340.

Andersen, T.G., T. Bollerslev, F.X. Diebold, and P. Labys. 2003. Modeling and Forecasting Realised Volatility. *Econometrica* 71 (2): 579–625.

Bandi, F., J.R. Russell, and Y. Zhu. 2008. Using high-frequency data in dynamic portfolio choice. *Econometric Reviews* 27: 163–198.

Carvalho, Marcelo C., Marco A. S. Freire, Marcelo C. Medeiros, and Leonardo R. Souza. 2006. Modeling and Forecasting the Volatility of Brazilian Asset Returns: A Realized Variance Approach. *Brazilian Review of Finance* 4: 321–343.

Chan, K. 1992. A further analysis of the lead-lag relationship between the cash market and the stock index futures market.'. *Review of Financial Studies* 5: 123–152.

Hasbrouck, J. 2003. Intraday price formation in US Equity markets.'. *Journal of Finance* 58: 2375–2399.

Ko, K., S. Lee, and J. Chung. 1995. Volatility, efficiency, and trading: Further evidence. *Journal of International Financial Management & Accounting* 6 (1): 26–42.

Madhavan, A., and V. Panchapagesan. 2000. Price discovery in auction markets: A look inside the black box. *Review of Financial Studies* 13 (3): 627–658.
Stoll, H.R., and R.E. Whaley. 1990. Stock market structure and volatility. *Review of Financial Studies* 3: 37–71.

Part II
Valuation

Valuation 4

Learning objectives:

- Begin enumerate types of valuation.
- Learn the models of value and distinguish among them
- Demonstrate the most uncertain present value in highly risky investments
- Highlight the importance of firm Multiples and their significance
- Describe accuracy tests of projected value, sampling, and comparison
- Learn Matching and mapping of models to purpose and characters
- Understand the relationship of firm value and impact of changes to volatility

4.1 Role of Valuation

Although valuators are widely recognized in the real estate industry, insurance industry, and among jewelers, stock valuation is also as interesting as others. Valuation is extremely useful in a wide spectrum across the industry. The role of valuation in portfolio exists in mergers and acquisition, equity holding, accounting and taxation (Modigliani and Miller 1958) purposes, etc. For instance, valuation plays a critical role in long-term horizon and fundamental analysis and short-term investing and relative valuation.

4.2 Types of Value

The types of values often highlighted by researchers and practitioners (Damodaran 1994) in the field cover the following as described below.

D. Bag, *Valuation and Volatility*,
https://doi.org/10.1007/978-981-16-1135-3_4

4.2.1 Fair Value

Fair value represents the price for possessing a title and the willingness of people to pay for it. The fair value of an object is taken into account for its benefit to utility and availability. When the benefits of possessing an object are widely known, the fair price increases assuming that goods are scarce. It is the intrinsic worth of the object or the true worth of the object. The investor demand for the stock determines the bid and asks prices, the exchange is a reliable method to determine a stock's fair value. The stock is trading at a fair value when there is consensus on the price, which remains in equilibrium. The consensus price may move up when all stakeholders believe that prices are unfair and a new consensus is reached.

4.2.2 Fundamental Value

Fundamental value is the accurately estimated present value of future cash flows. The future cash flows that accrue to an equity holder is the intrinsic value. The Intrinsic value is the worth of an asset that is derived independently of other extraneous factors. The subjective perception of a stock's value is reduced from the observed fundamentals and determining the worth of stock. The fundamental value of one stock is independent of other stock(s).

4.2.3 Terminal Value

Terminal value (TV) determines a company's value into perpetuity or the forecast period. When it is assumed that the firm ceases operations at a point in time in the future is called liquidation value. The liquidation value is based on the book value of the assets, adjusted for any inflation (deflation) before the end year. Asset valuation is undertaken to arrive at the replacement cost for firms that have assets that can be segregated and sold off. The liquidation value of a real estate empire can be derived from the selling off lock stock and barrel, or in parts.

4.2.4 Panic Value

Panic selling refers to the sudden, accidental, surprise sale of a stock caused by a sharp decline in price. Panic selling is escalated with investors' temptation to offload their holding. Panic selling is often triggered by an event that significantly decreases investor confidence in security. The fear leads to closing out positions prematurely. which is an irrational act to exit in haste? Fear morphs into a panic. Events during panic could be management changes, sales decline, falling revenue, drop in earnings, or top stocks biting the dust. A significant factor in panic selling is emotional trading. These trades are driven by fear, market sentiment, and overreaction to the news which could have real short-term effects. Most stock exchanges use trading halts to limit panic selling.

4.2.5 Resale Value

An owner of a liquid asset will be able to convert the asset to cash and will receive a value for the asset equal to the resale value. The current market value is the resale value of an asset. The resale value is the liquidity value of the stock. The current value is the closing price of stocks, or the last bid price offered over-the-counter (OTC). If the OTC market does not exist, e.g., resale value is hard to determine. The existence of a counter is necessary to obtain resale value. it is in smaller units the resale is possible than the aggregator who purchases the same from many sellers, with a discount.

4.2.6 Purchase Value

Purchase price is the initial purchase cost of the investment that equals the book value of an asset. The purchase price includes all admissible incidental costs incurred, viz., sales commission or charges paid to acquire the assets (inclusive of installation costs). The cost basis is the weighted average purchase cost for more than one purchase of the same stock at different points in time. For example, one buys 100 shares of ABC stock over a three-year period on three different dates. The weighted average purchase cost is a division of the total dollar amount and the number of purchased shares. The dollar amounts of a stock purchased at various points of time are $40, $60, and $80 for 100 shares each, which equals $18,000. The weighted average cost equals $60 per share as the cost basis. The net capital gains are the difference between sales and buy values, denoted as the average purchase price.

4.2.7 Worst Value

The maximum loss or possible drawdown is the largest price fall of an asset from a peak to a trough. In case of physical assets, the maximum foreseeable loss is the nominal value termed loss in insurance. At the trading desk, one can resist the temptation to sell off yielding to the grim situation to escape from realizing the worst value. The maximum foreseeable loss is the worst-case in which maximum damages could occur. In case of possessing physical assets, the maximum possible loss could be negative instead of zero. When the price falls below the book, or the price-to-book value is less than one (1), the bottom most is possible. Traders lose their margin to brokers. The traders forgo their balance of any cash or cash equivalent. Post the financial crisis in 2008, many stocks in the US had one below $1.00, however, they were maintained at the minimum of $1.00 to retain the respect of their globally branded stocks.

4.2.8 Best Value

The best value is defined as the most feasible balance between commercial benefits of an asset against its purchase cost. Best value for money is commonly known as the fruitful mix of cost, quality, and sustainability to meet customer requirements. The derived value is the perception of the worth of a product or service to a customer among its alternatives. Best price is a comparative picture obtained at the snapshot in time. For example, the best bid is the highest quoted price for stock among the bids displayed live in the exchange. The best bid refers to choosing scrip among many others in the same sector. The best bid is effectively the highest price that an investor would be paying for the asset in hand. Best value is comparatively identifying the monetary difference between a winning offer and other offers in hand. In case of a buy ordering system, the lowest price for a matching volume of orders is the most favorable bid instead of a bid with volume lower than quoted.

4.2.9 Exchange Value

It is the worth of one unit of good or service denominated in terms of the worth of another unit of good or service. Exchange rate is the denominated value of one currency expressed against another currency of another country. Exchange value differs from the "price" of an asset. Price is the *actualization* of exchange value or the number of goods and services in the exchange deal. The exchange value for a particular commodity varies from one market to another markets and it varies according to time and place. A given commodity may have immense use value but no exchange value or vice versa. For example, water has immense use value but not exchange value. On the contrary, diamond has huge exchange value but no use value. In a stock purchase, there exists the possibility of new stocks that could be exchanged against an existing stock or the conversion of the ownership of one stock with another stock in anticipation of an attractive offer by an acquirer.

4.2.10 Comparative Value

It is a tool that enables comparative assessment in the absence of computation. It provides a mechanism for valuing non-cash-flow-producing assets. This relies on algebraic manipulation and comparison of data among stocks rather than mathematical calculations. The multiples are numbers created from the ratio between a numerator and a denominator which are financials or operating ratios of the firm.

The peer-to-peer comparison is not easy. Examples of comparable peers include (1) size (revenue, total assets); (2) the rate of growth in earnings, (3) diversity of product ranges; (4) quality of the customer base; (5) level of borrowing, geography, etc.

4.2.11 Enterprise Value

Enterprise Value (EV) is the total value of a company. EV is the effective value of buying a company.

The formula is given as:

$$\text{EV} = (\text{Market Capitalization} + \text{Market Value of Debt}) \tag{4.1}$$

The value of equity is not independent of the value of the enterprise or entity, contingent claims or any other claims. The value of the firm accrues with active equity participation of the shareholders.

$$\text{EV} = (\text{Common Shares} + \text{Preferred Shares} + \text{Market Value of Debt} + \text{Minority Interest} - \text{Cash and equivalents}) \tag{4.2}$$

It is not easy to split and separate the value of equity and the value of the entity. Unlike a reserve of the gold mine which keeps depreciating over time and hence the value of the property, an enterprise that owns the gold mine does not depreciate over time. This is because the equity holders deploy enough capital to convert the raw reserves into productive commodities and so take full advantage of the future conditions in the commodity market.

4.2.12 OBS Value

OBS include the items off the balance sheet, such as guarantees, commitments, and positions in derivatives, etc. The company boards (BOD) may give guarantees in favor of subsidiaries or third parties during lending or business transactions. OBS exposures involve risks of market, operational, and credit risks, which affect the guarantors' solvency and liquidity. OBS activities positively influence a financial firm's stock return.

4.3 Value of Equity

4.3.1 Single Stock

Direct equity is the direct holding of the stocks of a company. The value of a single stock is important for individual owners or promotors, or insiders. The regulator provides alternate definitions of the offer price for intended acquisitions, depending on whether they are preference equity or stock options. During downslide, the company is vulnerable to takeover by creeping acquisitions by outsider owners, where a regulator defines the bid price of the offers. The transfer or acquisition of ownership is decided at a price, the higher of the negotiated price or, 52 weeks volume-weighted average price (VWAP) or 60 weeks volume-weighted average price (VWAP) prior to the date of announcement of the deal in public media.

$$VWAP_{60} = \sum_{i} \frac{V_{it}}{V_t} P_{it} \tag{4.3}$$

where $VWAP_{60}$ = VWAP for 60 trading days.

V Total turnover in the script on day t,

N_T Number of shares of the scrip traded on 'nth' trading day. NT is the sum total of the volume against each trade for the same scrip.

The single stock is held by an insider, where the insider is a director of a publicly-traded company, who owns more than a limit (10%) of a company's voting shares.

4.3.2 Stock Warrants

Warrants are allotted by companies to promoters, which could be based on the share price over the past 6 months. The value of warrants is cheaper than the market price. Warrants are sold at 25% of the value during the initial allotment. The balance 75% of the money is paid off when the warrant holder converts them into equity shares. Regulators require every listed firm to maintain a minimum public shareholding of 25%.

4.3.3 Employee Stock Options (ESOPs)

The ESOPs cost the employees at a price lower than the current market price. The ESOPs are purchased at the exercise price, which is mentioned in the offer. At times, employees are encouraged to purchase a small portion of the market price against matching contributions made by the employer. A higher stock price will benefit both the shareholders and the employees who own ESOPs. Insiders hold shares to acquire the company at a lower price and then benefit from selling them in the market at a higher price. Promoters of the company use ESOPs to own shares of the company at a low price. If promoters buy shares from the open market, the share prices will rise because the market perceives it as a positive signal. Promoters or insiders can acquire public issues in excess of the required minimum percentage, locked-in for a period of three years.

4.4 Valuation Approaches

Structurally, firms are group-owned, well-diversified entities or standalone business units (Haugen 1990). At the granularity of a division, they are denominated as service units or manufacturing units. Manufacturing firms own fixed assets to generate tangible output, maintain inventory, engage in labor-intensive operations and they need physical production facility and location. Compared to manufacturing entities, service units are associated with unique characteristics, such as intangibility, inseparability, heterogeneity, perishability, customer participation, etc. Because of the direct participation with the customer, the success of service operations is fundamentally dependent on knowledge and communication. Apart from higher fixed costs, the components of raw material and inventory costs as a part of the overall operating expenses of manufacturing units are higher than service. The manpower and communication costs as part of operating costs are higher fixed costs are lower in services than manufacturing. However, there are many parameters that are important for arriving at the value of the firm that is important to derive the changes to value (Kaplan and Roll 1972). The approach to value depends upon the nature, characteristics of the business, and its sources of cash flows.

The major numeric approaches to valuation include:

1. Income approach,
2. Discounted cash flow (DCF),
3. Dividend discount model (DDM),
4. Stable growth model (constant DDM),
5. Multi-state growth,
6. Relative valuation model,
7. Adjusted Valuation model.

The nature of valuation depends upon the purpose of whether it is equity value or firm value. The equity value is the fair value of pure legal ownership claims on the firm. Equity value is the net enterprise value of the firm exclusive of secured financial claims on the entity (McConnell and Muscarella 1985). Enterprise value (EV) is the grand total value of both equity claimants and non-equity financial claimants on the business (i.e., equity and debt holders.

Table 4.1 Examples of multiple comparisons

EBITA	Enterprise value	EV/EBITA
Earnings	Equity value	P/E (net income)
Book value	Equity value	P/B, price/*tangible book value*,
Revenue	Enterprise value	EV/Revenue

Multiples of the firm are used for valuing a business that has a definite stream of stable earnings. Book value multiples are used in industries where entities use their equity capital bases to generate earnings (for example, price/book value multiples for financial institutions). Revenue multiples are used for businesses that do not generate positive earnings (or loss-making). Table 4.1 shows examples of multiple comparisons.

(i) EBITDA and its multiple is the earnings (excluding items of adjustments to interest, depreciation and amortization (D and A), taxes from the earnings), which is useful for financial entities which lease their operating assets (i.e., less capital-intensive entities).
(ii) EBIT multiples denote adjusted depreciation and amortization. It reflects the replacement of assets. EBIT can vary depending on the existing presence and treatment or the amortization of intangible items. Ownership or acquisitions of brand or patents adds to amortized income. Higher EBIT denotes higher gross margin specific to a few sectors such as refineries or resellers or comparable units.
(iii) EBITA multiple is used in addition to EBIT when the level of intangible assets in a stock is lower than comparable peers. Higher EBITA denotes older firms with higher amortized pre-earned income.
(iv) Earnings are appropriate when all stocks have similar levels of tax payouts, earnings, and debt. This is used for financial firms (banking, insurance, and leasing) where the relevant operating expense is the interest expense. Trading on equity is observed in financial firms that tend to achieve higher gearing ratio.
(v) Book value is applied for comparing the market value with its book value of equity to identify undervalued or overvalued companies. This is not used for service firms because their intangible assets are higher as compared to fixed assets.
(vi) Revenue multiples are used if the earnings are highly correlated with its revenue. Multiples of revenue are applicable to service enterprises (e.g, advertising firms, professional practices, insurance agencies, etc.). Revenue multiples are also used when the entities are loss-making. Higher average price revenue per Unit (APRU) in telecom business. Number of subscribers, number of customers, to depict transparency.

Table 4.2 Cash flows and discount rate

Discounted cash flow models	Equity value	Enterprise value
Cash flows	FCFE are sourced from assets, after debt payments net of reinvestments are redeployed as cash	FCFF are sourced from assets, before any debt payments net of redeployed that are redeployed as cash
Discount rate	The discount rate (r) is the cost of equity (i.e. the *cost of equity capital, Ke*)	The discount rate is the effective cost of capital (i.e. the *weighted average cost of capital*, or *WACC*)

4.5 Models

4.5.1 Cash Flows (CF)

There are two types of cash flows (Bruner et al. 1998), namely, FCFF and FCFE, depending on end value or equity value, respectively (Table 4.2).

The discount rate is the weighted average cost of capital (WACC), proportionately weighted on both debt and equity.

WACC is given as;

$$WACC = \frac{E}{D+E}K_E + \frac{D}{D+E}(1-T)K_D \tag{4.4}$$

where D is the total value of debt funds;

E is the value of equity capital;
k_D is the effective cost of debt funds;
k_E is the effective cost of equity capital
t is the @ tax rate.

4.5.2 Cost of Equity Capital (K_E)

The cost of equity does not remain constant (Godfrey and Espinosa 1996). The changes to cost of equity (Ke) can occur because of changes to market expectations due to (1) outstanding equity, (2) buyback of shares, and (3) promoter holding. There are two choices, namely: (1) one factor model or (2) multi-factor model to calculate the cost of equity, respectively. The cost of equity (ke) is derived using the *capital asset pricing model (CAPM).*

The cost of equity (k_e) is given as:

$$k_E = r_F + (r_M - r_F)\beta \tag{4.5}$$

The cost of equity capital resulting from CAPM is an expected (market required) rate of return.

k_E is cost of equity (expected rate of return investors require on an equity investment);

r_F is risk-free rate (expected rate of return on a risk-free asset);

r_M is the required market rate of return (expected rate of return on a fully diversified portfolio);

$(r_M - r_F)$ is required equity premium in excess of the expected rate of return on a risk-free asset;

ß is measure of the systematic risk for the individual shares (i.e., the contribution to the variance of the market portfolio).

If the cost of capital is used to discount the cash flows to equity, instead of the cost of equity, the value of equity increases over its true value.

The second model uses a multi-factor APT equation that has more parameters. The arbitrage pricing theory (APT) model provides the cost of equity capital, where the expected returns increase linearly with a few number of factors, viz., inflation, GDP, etc. The cost of equity (Ke) in the APT equation, the expected return is given as.

$$K_E = \alpha + \beta_1 GDP + \beta_2 INFLATION + \beta_3 r_M + \varepsilon \qquad (4.6)$$

There is the advanced Fama–French (F–F) three-factor model that determines expected returns using three known factors: book-to-market factor, size factor, systemic factor, respectively.

4.5.3 Walter Model

The Walter model shows that the choice of dividend policies affects the value of the enterprise. The Walter model shows the relationship between the firm's internal rate of return (r) and its cost of capital (k) to maximize shareholders' wealth. Dividend models have the implicit assumption that the payout ratio does not change. The firm finances all its investment through retained earnings without any debt or new equity. Earnings are distributed as dividend or reinvested internally. The reinvested cash is discounted at the internal rate of return (IRR) and not at the ex-post invested cost of capital to calculate the net present value.

The discount model (DDM) provides the price of an entity's equity instrument (P0), which equals the present value of all of its expected future dividends in perpetuity.

$$P_0 = \sum_{0}^{\infty} \frac{D_t(1+g)}{(1+k_E)^t} \qquad (4.7)$$

where P_0 is the price of an equity instrument at time zero, Dt is the dividend to be received at the end of the period t, and k_E is the cost of equity capital.

$$\text{In perpetuity, model converges to } P_0 = 0\frac{D_0}{K_E - G} \tag{4.8}$$

4.5.4 Gordon Model

The dividend discount model (DDM) with constant growth assumptions generate the fair value from the available dividend stream. The dividends are projected into the future, while dividends grow at the constant growth rate (g).

If D_0 is the currently received dividend, expected future dividends are:

$$D_t = D_0(1 + g)t \tag{4.9}$$

The current price is given as:

$$P_0 = \frac{D_1}{K_E - G} \tag{4.10}$$

There are three inputs, expected dividends and the cost of equity (k_E), and growth in earnings (g) and payouts (%) that generate future dividends. The required rate of return (%) is derived from the CAPM model (e.g., β of the capital asset pricing model). The limitations are the dividends are not perpetually realized. The dividends depend on Dividend distribution tax, personal tax rates or corporate taxes (T), and shareholder activism, respectively. The limitation of the model is too sensitive to expected growth rates. The inputs to the model are payout ratio, cost of equity are assumed constant. Various assumptions of growth (g) include (1) no growth ($g = 0$), (ii) constant growth (g), and (iii) variable growth ($\delta g > 0$, $\delta g < 0$), respectively. The growth rate cannot exceed the cost of equity (Ke) since the piece of a share cannot be negative ($P < 0$). The Gordon model is suited for equity valuation done by outsiders for firms that are growing with dividend payout. Conversely, this model underestimates the value of the stock in firms that pay out less and accumulate cash.

4.5.5 H-Model

The H-model approximates the value of a company whose dividend growth rate is expected to change over time. It is the ratio between the Price determined at a low growth period and a price from the declining high-growth phase. It gives an alternate estimate of the constant growth. The growth in earnings is expected to decrease over time. The H-model gives a practical view of the declining growth rate, the current growth rate declines to reach the long-term growth rate. The shorter the high-growth period, H-model cannot forecast market share; instead, it is based on declining shares.

The half-life of the model (H) captures the rate at which the growth converges to a long-term growth rate. It is given as:

$$P_0 = \frac{D_0(1+g_s)}{r-g_L} + \frac{D_0 H(g_s - g_L)}{R-g_L}$$

where t is a period of high growth.

Where

D_0 is the current dividend,
g_S is short-term growth,
g_T is long-term growth,
r is required rate of return.

4.5.6 Grinold–Kroner Model

The Grinold and Kroner model is used to arrive at the forward estimates of equity-risk premium. The changes to outstanding equity due to stock repurchases will raise P/E ratio. The outstanding equity could fall due to the tax rate on dividend income, when the company may decides to buying back. Disbursal of stock dividends or bonus shares will reduce P/E. The expected return is related to the change in the P/E (ΔP/E) ratio. This is due to changes to outstanding equity.

Where D_1 is the dividend in next period, P_0 is the current price, π is the expected rate of inflation, g is the real growth rate in earnings, ΔN is the change in the number of shares outstanding, and (ΔP/E) is the change in the P/E ratio, respectively.

The nominal growth rate (g) is the sum of inflation (π) and real growth (g). The gordon growth model (GGM) is used to generate forward-looking estimates of the equity-risk premium. The risk premium is given as the expected growth rate (g) minus the current 10-year government bond yield (R_F). The change in the ΔP/E ratio positively influences the expected return.

$$E(R) = \frac{D}{P} - \Delta S + \Delta\left(\frac{P}{E}\right) + i + g$$

where D/P is the one-year forecast dividend yield on the market index,

g is the expected long-term earnings growth rate.

The limitation of GGM is the varying nature of equity-risk premium which impacts the calculated forward-looking P/E, and the model assumes constant growth is not practical.

4.5.7 Payout Multiplier Model (POR)

The pure dividend models assume that the payout ratio is never zero. If the initial dividend is zero, the payout ratio (POR = 0) is zero and the fair value needs to be determined. For example, the payout ratio could fall due to tax rate on dividend income, when the company decides to buying back. Further, the growth rate ($g \neq 0$) and it changes during the life of the firm. The POR model incorporates the market multiplier approach to find the dividend value of the firm in one or two stages, where the growth rates are assumed unequal among the economy and termination period of the firm.

The pure two-stage model determines the prices as below

$$P_0 = D_0 \frac{1+g}{d-g}\left[1 - \left(\frac{1+g}{1+d}\right)^T\right] + D_0\left(\frac{1+g_T}{d-g_T}\right)\left(\frac{1+g}{d-g}\right)^T$$

The P/E ratio is obtained from POR ($= \frac{D}{E}$) in the one-stage growth (g) model is given as:

$$\frac{P}{E_0} = POR\frac{1+g}{d-g}\left[1 - \left(\frac{1+g}{1+d}\right)^T\right] + POR\left(\frac{1+g_T}{d-g_T}\right)\left(\frac{1+g}{d-g}\right)^T \tag{4.11}$$

where,

d is the discount rate,
g is the growth rate,
g_T is the terminal growth rate.

Alternatively, the POR model is applied to two-stage determination of $\frac{P}{E_0}$ when the growth rate in the economy is g_M.

$$\frac{P}{E_0} = POR\frac{1+g}{d-g}\left[1 - \left(\frac{1+g}{1+d}\right)^T\right] + POR_M\left(\frac{1+g_M}{d-g_M}\right)\left(\frac{1+g}{d-g}\right)^t$$

where POR_M is the payout multiplier.

4.5.8 r-NPV (Risky NPV)

Discount rates are not known easily for the future and so are cash flows. The DCF formula can introduce the uncertainty in cash flows to arrive at the risk NPV. The r-NPV is a modification of the standard NPV method of discounted cash flow which considers the probability (P) of cash flow events. The numerator in r-NPV is the expected value of cash flows which is the product of projected cash flows and the probability of its occurrence. The accuracy of r-NPV is based on two more parameters than the basic NPV model.

The formula is given as:

$$rNPV_0 = \sum_{1} \frac{p_t C_t}{1 + d^t} \tag{4.12}$$

where C_t is the tth cash flow, p_t is the probability of meeting the projected cash flows from the initiation of the project and, d, is the discount rate.

The r-NPV approach reflects the relationship with the stages of a project under testing and is a better approach to value biopharma companies. To calculate the true present value of biotechnologies, revenue, cost, risk, and time must be combined into a single calculation of r-NPV. In the r-NPV equation, the present value of risk-adjusted costs is subtracted from the present value of the risk-adjusted payoff to arrive at the NPV of the biotechnology. It is applied to multistage product development and product launches are dependent on research and development costs. The approach could be implemented to project development and market launches when the inflows are specific to one or more scenarios. The limitations of DCF valuation lie in the fact that it does not reflect the micro situation of the market. The alternate r-NPV provides definite advantages, (2) it entails arriving at discount rates which need to be modeled rather than exogenously given.

4.5.9 Minority Ownership Discount

A minority discount is the opposite of a control premium. Control premium refers to the controlling value of shares. When the controlling shares in a public company are acquired, the buyer offers a premium to induce a large number of shareholders to sell their shares against the offer. The control premiums are derived from the sample data of non-controlling transactions on relevant acquisitions. Minority owners are not controlling owners. The voting rights of minority owners are below par the rights of promoter directors. The minority interest in a company's equity undergoing a valuation discount reflects the notion that a partial ownership interest may be worthless than its proportional share of the total business. The terms on whether an ordinary or special majority is needed and the power to appoint a director to the board, or the ability to control through such appointments. The nature of ownership control aims to influence both investment and operating decisions.

The discount for lack of control (DLOC) is determined from the control premiums offered to the existing shareholders during takeovers from past data. The control premia offered to shareholders depend on the geography, industry, and valuation multiple (P/E) of the comparable transactions in the sample of stocks. Previous studies (e.g., BCG) have shown that acquisition premiums averaged at 24.8% and varied between 15 to 40% in the sample. The control ownership of a 30% share in the business is lower than 30% of the total equity of the company. A discount of 20% on interest of 30% would reduce its effective worth to 24% of the total equity.

The minority discount (D) is derived from the control premium (P) as:

$$\text{Discount(D)} = \frac{1}{1+\text{Control premium(p)}} \tag{4.13}$$

4.5.10 Liquidity Impact

Liquidity is the fall in price when stocks are not traded for long. Does the liquidity discount depends on the nature of the stock? The discount is observed from the market prices of similar illiquid assets. The discount may vary across firms from relevant restricted pre-market stocks or with matching liquidity profiles of similar stocks in the sample of firms belonging to other markets. There exists a negative relation between turnover and illiquidity. The rates of discount may vary from 25 to 50%, with 35% as the median rate. The illiquidity discount is lower for larger firms. The rate of discount decreases when the earnings are positive. As discussed in Chap. 2, volatility and turnover exhibit a positive relationship between themselves. The final transaction prices from sample data of similar deals of liquidity are used. Liquidity discount is broken down into its components of brokerage or transaction costs. The reported "liquidity discounts" can reach up to 50% of value. The average liquidity discounts can range from 20–30%.

Therefore, multiples are chosen logically, and there exist limitations as shown in Table 4.3.

As shown in Table 4.4, the ABC company's current and quick ratios indicate the company's ability to pay its current payables. The firm ABC is stronger than the industry, although the ratios have declined over the years. This appears to be a result of decreasing inventory amounts. This is a positive sign because ABC was carrying older inventory, and a reduction in inventory will reduce the costs of carrying it. ABC's activity ratios appear to be fairly consistent from year to year. On average, it is collecting its receivables in less than 30 days, while the industry averages 30–45 days. Its inventory turnover has also improved, which is slower than the industry's. it is taking ABC longer to make payments on its accounts payable than in the past. This could be as a result of the large increase in revenues in the year 1994 than the year 1993, and the additional debt taken on over the last two years for the purchase of new equipment. The sales to working capital have been

Table 4.3 Limitations of multiples

Descriptions	Implications
– Inappropriate selection or wrong choice of comparable company peers	Pure earnings
– Mismatch between multiple chosen and historical earnings multiples on forward– looking earnings	Fundamental value
– Choosing post-tax numbers instead of pre-tax performance measures	Fair value
– Omission of adjustments that affect the valuation multiples while comparing company peers	Fundamental value
– Omission of adjustments for non–operating assets,	Fundamental value
– Discount for the illiquidity	Fair value
– Double-counting or omitting cash flows	FCCF
– Choosing wrong cash flows (FCFF or FCFE)	Fundamental value
– Discount rates, or cost of equity (WACC or KE)	Fundamental value
– Exorbitant growth rates assumption to arrive at the terminal value calculation	Fair value
– Actual life of the business	Fundamental value
– Deferred taxes and their omission	Fair value

stronger than the industry, although it is also declining as working capital declines. Overall, ABC appears to be operating efficiently and making good use of its assets.

4.6 Asset-Based Valuation Models

The asset value method comprises the traditional approach, common among accounting practitioners (Elton and Mei 1994). The asset value includes both replacement and liquidation values. In liquidation value, the assets are valued against similar assets available in the market. Relative valuation uses P/E, the price multiple for a firm. A higher P/E indicates better operating performance. Stocks with low P/E multiple compared to industry P/E. For example, many smaller names, viz., Escorts, REC, Hindalco, JK Tyres, Gujarat Narmada, Sun TV, etc., have higher prices. For example, a conservative method is when the earnings yield is higher than the yields of AAA bonds, the stock is undervalued. If E/P is lower than the AAA yields, the stock is overvalued. The dividend yield for dividend-paying stocks has a negative relationship with P/E multiples. At a sectoral level, P/E Multiples are negatively related to the index of industrial production (IIP). When the market is overpriced, any new entrants into the pre-market stocks or IPOs are overvalued.

Table 4.4 Comparison of multiples of ABC stock Ltd.

Ratio activity ratios	1999	2000	2001	2002	Analyst 1	Analyst 2	Analyst 3
Cost of sales to inventory	3.5	2.65	3.03	3.49	5.5	5.2	3.6
Days of inventory	104.39	137.88	120.65	104.58	66.36	78.6	75.4
Cost of sales to payables	14.87	13.02	14.76	10.19	13.02	14.76	10.19
Days of payables	24.55	28.03	24.72	35.82	28.03	24.72	35.82
Sales to working capital							
Operating ratios	3.76	3.65	3.75	4.41	8.9	8.3	8.3
Pretax profit to net worth	5.19	11.35	9.97	23.84	15.2	17.6	18.4
Pretax profit to total assets	4.56	10.07	8.17	17.76	5.8	10.5	9.8
Sales to net fixed assets	74.83	110.87	193.12	116.8	12.3	12.3	12.3
Sales to gross fixed assets	3.93	4.1	4.14	4.78	2.4	2.4	2.4
Sales to total assets	2	1.98	1.85	1.9	1.9	2.7	2.5
Coverage ratio (EBIT/Interest Expense)	12.03	44.92	16.37	31.76	3.3	3.3	3.3
Leverage ratios							
Net fixed assets to net worth	0.03	0.02	0.01	0.02	0.5	30.1	30.1
Total debt to net worth	0.14	0.13	0.22	0.34	1.6	62.1	111.8
Liquidity ratios	5.79	5.81	5.4	3.72	1.7	2.3	2.6
Current ratio	2.99	2.36	2.33	1.78	0.8	1.2	1.6
Sales to receivables	13.05	13.18	15.03	14.22	8.7	10.52	7.9

Firms depict a differential response to the economy, such as inflation or interest costs. CAPE ratio is an adjustment of inflation to earnings multiples. The price-to-earnings ratio (CAPE) or Shiller P/E is cyclically adjusted earnings. It ignores the shorter one-year earnings; instead, Shiller's P/E ratio is the average of the last 10 years the earnings adjusted for inflation. The ratio of the average P/E to the current index is the CAPE ratio. When the CAPE ratio is above the stock's historical average the market is "overvalued". For example, when the CAPE ratio is higher than its long-term average, the market is "undervalued". Automobile stocks undergo swings in earnings. The company's P/E ratio falls. This would suggest that the stock has become cheaper and the P/E ratio will increase.

4.6.1 Terminal Liquidation and Replacement Value

The liquidation value at the end of a fixed period of time is equal to the terminal value (TV). Companies reinvest their surplus cash into new assets to extend their lives. Beyond the terminal year, the cash flow grows at a constant rate to perpetuity. Terminal value arrives with the dividends (D), or free cash flow (FCFF) forecasted for a fixed period. Subsequently, the firm is expected to remain active for an infinite period. It is difficult to determine the finite horizon after which the firm could cease operations. Older firms, that are matured, tend to grow at a slow rate. An ideal three-stage growth model can possibly capture the long-term transitions to stable growth. Firm's growth is never independent of the parent owner entity. Growth depends upon stepwise future investments in fixed assets. The service sector may grow faster than manufacturing firms. Growth is higher when there exists access to supported debt infusion. Product growth reaches stability earlier than the firms' growth. The realized growth also depends on the ambition of the management owners of the firm.

The following is the formula to calculate the terminal value

$$TV = \frac{\text{FCFF}(1+\text{g})}{d-g} \tag{4.14}$$

.

where,

FCFF ree cash flow for the last forecast period.
g Terminal growth rate.
d discount rate.

The growth rate (g) is closer to the rate of inflation (π), and is lower than the GDP growth rate (G).

The choice of discount rate (%) depends on the cash flows of the firm versus the cash flows to equity are to be used. Two principals, namely: (1) The FCFF calculates the value of the firm, and (2) FCFE calculates the value of equity, respectively. The difference between both the cash flows is the magnitude of capital expenses. The FCFF amount is preferred for firm valuation since it is less sensitive to assumptions about growth and risk than FCCE.

(1) FCFF is expressed as follows:

$$FCFF = (EBIT(1-T) + \text{Depreciation and Amortization (DA) - CapEx} \\ - \Delta\text{Working Capital}) \tag{4.15}$$

where,

EBIT(1–T) is the net income (@tax rate T),

Depreciation and amortization (DA) are the Non-cash charges,

CapEx is the Capital Expenditure,

ΔWorking Capital is the changes to net Working Capital.

The cash flow to equity equals dividends or a free cash flow to equity.

(2) FCCE is expressed as follows:

$$FCFE = (EBIT \; - Interest\ Expenses - Taxes - \Delta Working\ Capital \\ - CapEx + \Delta Net\ Borrowing) \tag{4.16}$$

where.

CapEx is the Capital Expenditure,

ΔWorking Capital is the changes to net Working Capital,

Net Borrowing (ΔD) is the addition to net debt.

4.6.2 Leverage Buyout (LBO)

A leveraged buyout (LBO) is a channel determining the levered value of a firm. The raised debt would exist in the name of the newly formed entity. A financial buyer (e.g., private equity fund) invests a small amount of equity (relative to the total purchase price) and uses leverage to fund the total cost of acquisition to the shareholder. The LBO channel provides the "floor" value of the standalone company. The leverage level leads to arrival at the multiple at which the buyer is expected to exit the target. LBO investment denotes "cash-on-cash" (CoC), which is the final value of the firm at exit divided by the value of initial equity infusion. LBO ownerships may generate returns that equal 2.0–5.0 times the cash-on-cash (COC) (Table 4.5).

4.6.3 P/E Ratio

The P/E ratio indicates how expensive a stock is. This ratio of the share price (P) and earnings per share (EPS) measures how many times investors pay the current earnings. A low P/E ratio firm is expected to earn its investment back faster than a higher P/E ratio firm. A P/E ratio below 10 is cheaper. P/E ratios higher than 20 are expensive. The price earnings ratios are relative to geography, sectors, and time, etc. In developed markets, financials trade higher than multiples of 2.0, whereas in emerging markets, they trade at multiples exceeding 8.0. For example, when USD is higher, it is quite normal for exporters to earn more than non-exporters. When USD falls back to normal, quickly, the earnings of exporters are as much as non-exporters. P/E is suitable for financial service firms. For oil refineries, crude price rise can improve margins until the crude prices fall back to normal.

Table 4.5 Cash-on-cash (COC) and exit multiples for LBO for ABC Ltd.

Year	1	2	3	4	5	6	Total
EBITDA ($M)	40	44	48	53	59	64	
FCCF ($M)	5	6	7	8	9		35
ENTRY					*EXIT*		
MULTIPLE	5				MULTIPLE	5	
EBITDA($M)	40				EBITDA($M)	320	
PRICE PURCHASE ($)	200						
DISCOUNT RATE (%)	10				Total FCCF($M)	35	
DEBT($M)	120				DEBT($M)	85	
EQUITY($M)	80				EQUITY($m)	235	
					ENTERPRISE VALUE	3	

4.6.4 Adjustments to Price

Stock prices are adjusted in situations when the company announces stock splits, dividend payouts, and rights or bonus offerings, etc., it affects a stock's price since the company's number of outstanding shares increase by a factor of three (3.0), If a stock closed at $150 the day before its stock split, the closing price is adjusted to $50 (1: 3 ratio) per share to make for the change.

Dividends payments by a stock include cash dividends and stock dividends. Suppose a company declared a $2.0 cash dividend and is trading at $55 per share on the ex-dividend date. On the ex-dividend date, the stock price is reduced by $2.0 to report the adjusted closing price at $53.

Rights Offers

The closing price is adjusted to reflect rights offers extended to shareholders. The rights offer are rights issued to existing shareholders to subscribe to the new issue in proportion to their shares. The price of existing shares falls with the rise in the number of outstanding equity shares. The supply shock will reduce prices where the allotments are in a fixed ratio of 1:1, 2:1, 3:1, etc. If the ratio is 2:1, for each share, the existing shareholders get two new shares. When the bonus shares are issued, the number of shares will increase outstanding equity against the original purchase value of an investment that remains unchanged.

EPS Bootstrapping

EPS bootstrapping is used to enhance the EPS of a firm to boost the stock price. The merger between two stocks A and B is shown to increase projected earnings per share. In bootstrapping, the acquirer buys a company (B) that has a lower price earnings (P/E). The goal of the merger is to boost the post-acquisition EPS of the newly created company and increase the stock price. The EPS improves automatically when the acquirer's P/E is higher than the P/E of the target company.

4.6.5 Lack of Marketability (LOM)

It is practical to apply a discount for the Lack of Marketability (DLOM) for illiquid stocks. At one extreme, there lie unlisted stocks that have no liquidity. The discount on the stock for its lack of marketability is due to four major approaches. The first approach uses the price of restricted stocks compared to the price of listed and traded stocks. The second approach uses the ratio between the price of an ATM (at the money) put option and its underlying spot price. The distance between the put option prices relative to its spot indicates the degree of illiquidity. The third approach compares the ratio between the prices of pre-IPO shares of the company to the price of the post-IPO shares. The post-IPO shares have a lower price risk compared to pre-IPO shares.

The stock (restricted) approaches to unregistered shares of ownership in a corporationare issued to promoters, insiders, executives, and directors, etc. The restricted stocks are not transferable and are traded in compliance with securities and exchange commission regulations. The restrictions are intended to prevent premature selling from preventing adverse impacts on the company. The stockholders pay tax on capital gain arrived by the difference in the price of vesting and selling. An employee pays income tax on the total value of the stock during the period it vests. The employee also pays capital gains tax on any gains in the value of the stock when it is sold. The tax incidence on the owner depends upon the difference between the market value of the stock and the original exercise price. The discount for liquidity tends to be smaller for entities with higher revenues.

4.7 Choices in Models

The analyst faces the challenge of valuing a firm driven by the characteristics of the firm/asset being valued. Matching the valuation model to the assets of the firm is needed. In case of multiples, the equity value or firm value is related to firm-specific variables, earnings, book value, sales, etc. The multiples are arrived at from a sample of comparable firms in a similar business, where the comparable in the same geography has similar characteristics.

4.7.1 Matching Methods

Table 4.6 shows the mapping of the methods to the Context of valuation.

Business is understood as a whole and not in parts. Partnership value of a business is more than its standalone value (Estep 1987). The projected values

Table 4.6 Mapping the methods to context

Name of the model	Dividend	Life	After Life	Growth (g)	Context
1. Walter	0	Fixed	Finite	> 0	Fundamental Value
2. Gordon	> 0	Fixed	Finite	> 0	Fundamental Value
3. Two Stage	D1 > 0, D2 > 0	Fixed	Finite	> 0	Fundamental Value
4. Alternate growth	gT > 0	Fixed	Finite	> 0	Fundamental Value
5. Exit Multiple (s)	–	Fixed	–	> 0	Relative Value
6. Terminal Value	–	Fixed	–	> 0	Liquidation Value
7. LBO	–	Fixed	–	> 0	Levered Buy
8. Others	–	Fixed	–	> 0	Option Value

obtained from the above approaches can vary widely from each other. For example, in service entities, one may include lower fixed assets, more leased assets, leverage, ratio of fee to non-fee income, and the fee to non-fee expenses, etc. The brand value is inherent in the business itself. Similarly, for a FMCG major, the razor business cannot be segregated from the shaving cream business and other similar units. The company's assets are classified into cash-generating abilities, namely: (1) current assets, (2) future assets or, (3) non-cash rights, respectively. The first group of current assets is subject to fundamental valuation. The second group of future assets is R and D and knowledge assets, which could generate cash such as patents, technology, oil or mining reserves, and land banks. These assets could generate cash flows in the future only when the fixed costs are incurred. The staggered fixed costs are handled as an option as in binary choice models such as option pricing models. Lastly, assets that do not generate cash include cash rights, such as a club membership or fine art, etc., which can only be determined using replacement costs or relative value models.

There may exist outliers firms with no direct or relevant comparable, with little or no revenues, and with negative earnings. The sample selection is an art that requires skill and experience. In fairly emerging sectors, the sample should include similar numbers available for relevant sectors from other continents or geographies. When firms have unstable leverage, the firm value (the outstanding debt is subtracted) is appropriate to use. The average costs of capital (WACC) for low-levered firms are approximately equal to their cost of equity.

When earnings are negative or abnormal, the current earnings are replaced by a normalized value obtained from historical data or industry averages. Firms can have extended periods of negative or low earnings. The base year or normal year from historical data is chosen so as to reflect the most normal circumstances of the business cycle for the firm. The normal year is not a year of management fiasco, regulatory dictate, export debacles, disasters, gloom period, etc.

4.7.2 Choice of Multiples

The major types of multiples are based on (1) earnings, (2) book value, (3) revenues, etc. Apart from the available classes of multiples, a practitioner may attempt to use newer innovative multiples. The choice of a particular multiple can make a huge difference to the estimate (Dimson 1979). The choice of right multiple is undertaken by comparative assessment rather than individual bias. The other approach is the weighted average, where the weight is assigned on each indicator multiple. The financial profiles of firms are supplemented to choose the right multiple. When the number of multiples is in dozens or more a ranking scheme is used. The final ranks have arrived from each of the indicators in the matrix. For example, post COVID-19, the differential growth rates that could be observed across firms and P/E ratios are less accurate. India is considered a growth market and the multiples are expensive compared to Europe. The accounting earnings would vary on the choice of depreciation method. Cash flows are better measures than equity earnings if there are significant depreciation charges. The price-to-book (P/ BV) ratios are the most commonly used to analyze financial service companies. The price-to-book ratios are used when market ratios are less reliable. Table 4.7 shows the choice of Multiples by stock Sector.

In the sample data of multiples, an average or median of the multiples is selected when the investee characteristics are closer to comparable company peers. The minimum or maximum values of the multiples are used depending on whether the target is underperforming or overperforming. The multiples at the maximum of the range are mapped to those of comparable better-performing multiples. The value of multiple at the minimum of the range is mapped to those of comparable poor-performing peers. Between the average and median values of the multiples, the average values are preferred for large samples or multiples which give wider distribution in values, viz., ratios. However, for ratios using book value or its form in the numerator, the mode is preferred than mean values because they are discrete in nature.

The relative valuations determine whether firms in the business (sector) are undervalued or overvalued. A firm can be undervalued relative to its sector but overvalued relative to the market. If the stock is undervalued in relative valuation and appears overvalued in DCF models, the entire sector is overvalued. When the stock is undervalued relative to peer firms. the sector might be overvalued compared to all sectors. If a firm is undervalued in the DCF model and overvalued using the relative valuation, it indicates that the stock sector is undervalued.

Table 4.7 Choice of multiples by stock sector

Sector	Multiple(s)
Cyclical manufacturing	Relative P/E
High tech, high growth	Relative PE = PE of Firm / PE of Market. The same measure of earnings for both the firm and the market is used in numerator and denominator
	P/E growth = *P/E / g*
	P/EG ratio is a company's Price/Earnings ratio divided by its earnings growth (g) over a period of time
Infrastructure High growth, Negative earnings	P/EBITDA = M.V / EBITDA
	EV/EBITDA ratio compares a company's enterprise value to its earnings before interest, taxes, depreciation, and amortization
Real Estate	P/CF
	Value of a stock's price relative to its operating cash flow per share
Financial services	P/BVPS
Retailing, FMCG	P/BVPS
	The PBV ratio is the market price per share divided by the book value per share

Note These illustrations are exemplary in nature. The lower ranges and upper ranges are more important

4.7.3 Control Premium

The premium for control is the difference between the price before and after the transaction is equal to the premium paid for the controlling interest in the stock. The average of such transactions gives rise to the average premium paid. The discount for non-controlling interest is applied to the valuation multiples of sample stocks selected by matching the profiles of firm size, vintage, sector, geography, country, industry, period, regulatory environment, etc. The average of the sample data of transactions gives rise to the average discount to be applied. The matched micro parameters include the situation of the stock markets, leverage of the acquirer, leverage of the targets, management experience, payments modes, etc. The limitations of post-merger prices of stocks lie in valuation multiples from prior transactions. The prior transactions are the mix of control premium and synergy premium, which cannot be separated from each other. The prices of post-merger stocks in the sample are noisy, and hence pre-merger price is better to use.

4.8 Impact of Volatility

Businesses are impacted by volatility or risks of the nature of market risk, credit risk, operation risk, etc. Volatility and market prices tend to exhibit an inverse relationship. Volatility will impact the cost of borrowing from the market and stock liquidity. Uncertainties in cash induce volatility flows from long-term assets and short-term liabilities (Hamada 1972). Volatility discourages insurers and pension funds to hold positions in long-term assets. The response of volatility to spot prices of stocks occurs with a lag because volatility persists. The relationship between volatility is described in Table 4.8 given below.

The relation between volatility and the realized risk premium is negative ex-post (Fuller and Hsia 1984). The relation between risk premiums and volatility is positive ex-ante. The relationship is positive for firm-level variables, viz., book value of the equity (BVE), gross earnings (earnings before extraordinary items and depreciation) (EBITDA), sales revenue growth, return on equity (ROE), gross profit margin, etc. As mentioned in Chap. 2, idiosyncratic volatility is the unexplained variance in stock price not attributed to systemic risk. The bigger entities have lower idiosyncratic volatility, higher Interest coverage, high ROE, high EPS, and

Table 4.8 Relationship between stock volatility and its determinants

Characteristics	Nature of relationship	Remarks
Stock price	(–) negative	Prices fall with volatility
Expected risk premium (ex-post)	+ positive	Expectations increase with rising volatility due to risk-averse behaviur
Expected risk premium (ex-ante)	(–) negative	Negative because volatility induces lower returns
Book value of equity (BE)	+ positive	Purchase of undervalued stocks gives rise to high volatility
Market capitalization	(–) negative	Market price falls with volatility
Sale growth	+ positive	Good news for the market, and it responds
Return on Equity (ROE)	+ positive	Higher return for the market and it responds
EPS	+ positive	News of earnings for the market and it responds
P/E	+ positive	Price falls more than earnings. Good news for the market, and it responds
EBITDA	(–) negative	Earnings improve the liquidity of the stock
Gross Margin	+ positive	Margins improve the liquidity of the stock
Return on Stock	+ positive	Returns improve the liquidity of the stock
Interest Coverage	(–) negative	Coverage downplays the volatility effect of existing debt

high PE. The smaller companies tend to have higher idiosyncratic volatility, lower interest coverage, low ROE, low EPS, and low PE. The relationship with idiosyncratic volatility is not the same. An inverse relationship between the volatility and liquidity of assets. Volatility affects the market risk premium since a higher discount rate reduces the equilibrium price. The volatility of stock causes risk premium to rise, leading to a positive correlation between volatility and returns. The higher risk induced by the business increases the discount rate when the fixed earnings and dividends are translated into the current market prices. This results in a lower price for a given forecast level of earnings. Higher inflation reduces the allowance for depreciation relative to actual economic depreciation, which results in lower price/earnings multiples. The exchange rates cause a positive impact on earnings.

For financial service firms, the relationship between their exposures of Risk-Weighted Assets and market risk is negative. This reflects that the volatility of a bank's stock rises with increasing RWA, non-performing assets (NPA) projections and asset growth, etc. OBS exposures have a negative impact on return on equity. The negative relationship is due to the expectation that OBS activities reduce the risk exposure of the banks. A positive relationship between volatility and risk exists because the firm's guarantees do not appear on the balance sheet. Overall, OBS items reduce the total risk of firms.

Unsystematic (Idiosyncratic) Risk

Idiosyncratic volatility is the specific risk which is the complement of systemic risk. It is the complement of systemic risk. Returns are negatively related to idiosyncratic risk. The positive relationship between the returns and idiosyncratic risk holds for daily, weekly, monthly returns for sampling windows for one month, two months, three months, etc. When the sample is controlled for firm size, a positive relationship exists between dividend yield and idiosyncratic volatility, a negative relationship between ROE and idiosyncratic volatility, and a negative relationship exists between P/E ratio and idiosyncratic volatility.

4.9 Tests of Robustness

The practitioner's test of the model is desirable (Rosenberg and Marathe 1979). It is a test to validate the rank ordering ability of the valuation model used. The test proceeds by forming ten (10) equally distributed deciles of stocks ranked by descending values of DCF. Later, the descending values of mean projected DCF values and the averages of major financials of the stocks in the groups are observed. This is to validate the rank ordering ability of the test metric of projected values. The analyst could add additional profiling attributes to enrich and supplement the analysis as a confirmatory test of the results. There is no guarantee that the accuracy will improve with longer vintage or more liquid stocks in the sample.

4.9.1 Sampling Plans

The sample period depends on the volatility of the stock, which is normally higher for big stocks. The sample size is 200 days or more of traded observations days for each stock. However, in case of portfolio tests, the sample size is observed in the number of days closer to 100 days. The split sample test of projected value comprises sample cohorts of stocks. A larger sample size of stocks is desirable to conduct split-sample validation tests to compare the actual prices with computed values. For any given stocks, consecutive days of observation are used as sample day observation records. The sample windows are split into subsamples or smaller windows. The estimate in each subsample is almost equal to that of the full sample or close enough. The cohort pair (s) can be diagnosed for observed differences with a paired t-test (or between comparison tests) using standard t-test techniques. The error is the difference between the current price and the projected price.

The evaluation is based on any given group of securities when the securities are further broken down into their size, volume, and volatility characteristics, respectively, as shown in Table 4.9.

As shown in Table 4.9, a total of 12 groups are formed containing the size, volume, and volatility. The errors could be higher with higher volatility-induced discount rates. A given stock falls into one of the complimentary combination groups of low volatility, low volume, and large stocks. The errors are compared for each model to choose the appropriate. The comparison across stocks gives an overall picture of the robustness of the performance.

The full sample is used to divide data into in-sample and hold out sample. The split point can lead to distortions minimized by selecting the split point in the beginning or midpoint of the sample. Alternatively, a simultaneous test for out-of-sample accuracy at multiple split points is suggested.

4.9.2 Errors in Value

There are two types of errors, pricing error and valuation error, respectively. Pricing errors and valuation errors are independent of each other. The pricing accuracy of fundamental value is determined by observed differences between model values and stock market prices at a future date.

Let V_0 = projected value at $t = 0$,

P_0 = market price of owners' equity at $t = 0$,

An accurate pricing model ensures that the values are unbiased when an error (ε) is close to zero. The random term ε is market efficiency.

$$P_0 = (V_0 + \epsilon) \tag{4.17}$$

where, σ_e^2 is the variance of the pricing error.

σ_e^2 denotes the standard deviation of the pricing error ε.

Table 4.9 Error (s) in projected value by stock groups

Stock name	Size	Volume	Volatility	Error in projected Value (%)
A	Small	Low	Low	10
B	Medium	Medium	Medium	10
C	Large	High	High	15
D	Small	Low	Low	10
E	Medium	Medium	Medium	5
F	Large	High	High	10
G	Small	Low	Low	15
H	Medium	Medium	Medium	15
I	Large	High	High	10
J	Small	Low	Low	10
K	Medium	Medium	Medium	10

Table 4.10 Two way error

		Actual	
		1	0
Predicted	1	85	15
	0	15	85

Note 1—Over valued, 0—Undervalued

The forecast evaluation tests can lead to spurious evidence of predictability. The forecast of stock price (P) is compared across models.

While comparing between models, the common hypotheses are incorporated as follows:

1. The standard deviation of Errors (σ_e^2) for each model is equal.
2. Buy signals of each model are strongly positive ($V_0 - P_0 > 0$).
3. The mispricing ($V_0 – P_0$) of any two models are the same.

There exists a tradeoff between the errors generated from one model and the other. In this context, the accuracy in such models is improved when the total errors are reported, as shown in Table 4.10.

4.10 Fitness Tests

GOF test is a measure of the degree of coherence between the actual and projected identifiers. The goodness of fit test is a two-way test to assess the actual prices with projected prices in the sample data. The GOF tests for larger samples are conducted in sample records in equal sizes of ten (10) groups each.

Table 4.11 Goodness of Fit

Probability	Number of observations	Total number of observations	Proportion
Probability (Value actual > Value predicted)	118	250	0.472
Probability (Value actual > Value predicted)	32	250	0.128 =
Probability (Value actual > Value predicted)	100	250	0.4
Measure	Indicator	Mean (Indicator)	Remarks
Value actual > value predicted	1	0.55	Significant
Value actual > value predicted	–1	–0.45	Significant
Value actual > value predicted	0	0.002	Insignificant

The calculated prices from the model are transformed into numeric indicators of + 1, 0, or –1, respectively. For example, the following Table 4.11 explains the transformation.

A nonparametric test of the difference between actual and predicted is more feasible.

4.11 Summary

This chapter discussed major approaches to valuation, methods, and their applicability to make better decisions. It described major models of valuation. It summarised the DCF discounted cash flows to generate the present value. The relative valuation attempted to link between multiples and fundamentals. It tabulated the mapping and the manner to choose and apply these models. It explained the relationship between volatility and firm characteristics of the stock. It is practical to realize that the markets may be overvalued in one geography compared to other merging sectors. It discussed the backtesting of fundamental value, the sampling plan, and test methods for accuracy tests. It narrated the principal determinants of changes and adjustments to value due to liquidity, lack of ownership, etc.

4.12 Questions

4.12.1 Calculate the fair price of a Pharma stock using the r-NPV method (Table 4.12)?

Table 4.12 Cash flows for pharma company

Year	Description	Nominal cost	Probability
1	Phase 2 prototype1 costs	−660	100
1	Phase 2 prototype2 costs	−500	100
2	Phase 2 animal trial type 1 costs	−660	100
2	Phase 2 animal trial type 2 costs	−2200	50
3	Phase 2 human trial type 1 costs	−375	48
3	Phase 2 human trial type 2 costs	−2200	49
4	FDA Filing Costs	650	43
5	Royalty and Net Margin from Sales−Launch	75,000	35
6	Royalty and Net Margin from Sales-Launch	150,000	35
7	Royalty and Net Margin from Sales- Normal	225,000	35
8	Royalty and Net Margin from Sales-Normal	225,000	35
9	Royalty and Net Margin from Sales-Normal	225,000	35
10	Royalty and Net Margin from Sales-Normal	225,000	35
11	Royalty and Net Margin from Sales-Normal	225,000	35
12	Royalty and Net Margin from Sales-Normal	225,000	35
13	Royalty and Net Margin from Sales-Normal	225,000	35
14	Royalty and Net Margin from Sales-Normal	225,000	35
15	Royalty and Net Margin from Sales-Decline	150,000	35
16	Royalty and Net Margin from Sales-Decline	75,000	35

4.12.2 Calculate the stock price at the end of (i) 3rd year and (ii) 15th year for ABC stock using constant dividend growth rates in the model is @4% when D_0 = $1.95 and R = 10.5%?

4.12.3 If company A has a higher return on Equity (ROE) than company B, does it mean that the company A's earnings grow higher than Company B's earnings. Why?

4.12.4 If Dividend decisions are irrelevant to the value of a firm, then why use the DDM model at all?

4.12.5 Calculate the price of the stock from the data if the stock pays no dividends for 10 years. The stock begins paying dividends after the tenth year. It will have a constant growth rate of dividends.

4.12.6 Calculate the price today if the dividends are declining ($\Delta D < 0$) at g (=3%)?

4.12.7 Calculate the target stock price from the following information? Calculate dividend growth, g?
Net income = $875,000, Equity ROE = $7,300,000,
Shares outstanding = 125,000, Dividend = $345, 000, P/E ratio = 16.0.

4.12.8 When is the dividend discount model rather than a free cash flow model was chosen to value a firm?

4.12.9 When is the multistage dividend discount model is more appropriate than the constant growth models?

4.12.10 If a security is underpriced (i.e., intrinsic value > price), what is the relationship between its market capitalization rate and its expected rate of return?

4.12.11 The risk-free rate of return is 8%, the expected rate of return on the market portfolio is 15%, and the beta of the stock is 1.2. The company pays out 40% of its earnings in dividends when the latest earnings announced were $10 per share. One expects the company to earn an ROE of 20% per year on all reinvested earnings forever.

a. What is the intrinsic value of a share of stock?
b. If the market price of a share is currently $100, and you expect the expected growth in earnings g to be 12% and the market price to be equal to the intrinsic value of 102 in one year from now, what is your expected 1-year holding-period return on the stock?

Solutions

4.12.1 See Table 4.13.

The r-NPV for row 3 is calculated using the formula:

$rNPV_0 = \sum_1 \frac{p_t C_t}{(1+d)^t}$ = (–660 * 100/100)/(1.15 ^ 1) = −570.

The r-NPV for row 14, is calculated using the above formula equals.

= (–2,25,000 × 35/100)/(1.15^14) = -11,130.

The total r-NPV, at a discount rate of 15%, is $160,270.

4.12.2 Calculate the stock price at the end of (i) 3rd year & (ii) 15th year for ABC stock using constant dividend growth rates in the model is @4% the model is @4% when D0 = $1.95 & R = 10.5%?

$$P_t = \frac{D_t(1+g)}{(R-g)}$$

the current price of the stock $P_0 = D_0\ (1 + g) / (R - g)$ = $1.95 (1.04) / (0.105 – 0.04) = $31.20.

The dividend at Year 4 is the dividend D_0 for the growth rate in a four year period.

$P_3 = D_3\ (1 + g) / (R - g) = D_0\ (1+g)^4 / (R - g)$ = $1.95 $(1.04)^4$ / (.105 – .04) = $35.10

Or alternatively, $P_3 = P_0(1 + g)^3$ = 31.20(1 + .04)^3$ = $35.10.

the stock price at the end of 15th year is

$P_{15} = P_0(1 + g)^{15}$ = 31.20(1 + .04)^{15}$ = $56.19.

Table 4.13 r-NPV for pharma company

Year	Description	Nominal cost	Probability	Risk-adjusted income (loss)	r-NPV
1	Phase 2 prototype1 costs	−660	100	−660	−660
1	Phase 2 prototype2 costs	−500	100	−500	−500
2	Phase 2 animal trial type 1 costs	−660	100	−660	−570
2	Phase 2 animal trial type 2 costs	−2200	50	−1070	−810
3	Phase 2 human trial type 1 costs	−375	48	−180	–140
3	Phase 2 human trial type 2 costs	−2200	49	−1070	
4	FDA Filing Costs	650	43	−1070	–700
5	Royalty and Net Margin from Sales−Launch	75,000	35	−280	120
6	Royalty and Net Margin from Sales-Launch	150,000	35	26,250	11,350
7	Royalty and Net Margin from Sales- Normal	225,000	35	52,250	19,740
8	Royalty and Net Margin from Sales-Normal	225,000	35	26,250	25,740
9	Royalty and Net Margin from Sales-Normal	225,000	35	52,500	22,390
10	Royalty and Net Margin from Sales-Normal	225,000	35	78,750	19,470
11	Royalty and Net Margin from Sales-Normal	225,000	35	78,750	16,930
12	Royalty and Net Margin from Sales-Normal	225,000	35	78,750	14,720
13	Royalty and Net Margin from Sales-Normal	225,000	35	78,750	12,800
14	Royalty and Net Margin from Sales-Normal	225,000	35	78,750	11,130
15	Royalty and Net Margin from Sales-Decline	150,000	35	52,500	6450
16	Royalty and Net Margin from Sales-Decline	75,000	35	26,250	2810
			Total		160,270

4.12.3 No. Although growth is related to payout ratio, it varies across companies. It is not true.

4.12.4 If Dividend decisions are irrelevant to the value of a firm, then why use the DDM model at all?

DDM is applied to stocks that do not pay dividends after arriving at expected dividends that could have been paid if they were declared. Modigliani's hypothesis did mentioned that dividend decisions were irrelevant to arrive at a firm value.

4.12.5 The formula is

$$P_t = \frac{D_t(1+g)}{(R-g)}$$

the stock price at the end of 10th years equals,
$P_{11} = P_0(1 + g)^{10} = \$31.20(1 + .04)^{11} = \$56.19$

4.12.6 Calculate the price today if the dividends are declining ($\Delta D < 0$) at g (=3%)?
The price of the stock today is

$$P_0 = \frac{D_o(1+g)}{(R-g)}$$

$P_0 = \$10.25(1 - 0.03) / [(0.095 - (-0.03)] = \$79.54$

4.12.7 The growth rate g is due to the retention ratio
ROE = Net income / Equity ROE = \$875,000 / \$7,300,000 = 0.1199, or 11.99%
the retention ratio, is one minus the payout ratio, or

$$\text{b} = 1 - \left(\frac{\text{Dividends}}{\text{Net income}}\right)$$

= 1 – \$345,000 / \$875,000 = .6057, or 60.57%.
= (0.1199 × 0.6057) / (1 – 0.1199 × 0.6057) = 0783, or 7.83%

$$g = \frac{\text{ROE} \times \text{b}}{1 - \text{ROE} \times \text{b}}$$

The current

$$\text{EPS}_0 = \frac{\text{Net income}}{\text{Shares outstanding}}$$

EPS_0 = \$875,000 / 125,000 = \$7.00
The EPS next year $EPS_1 = EPS_0(1+g)$= \$7.00(1 + 0.0783) = \$7.55.
The target share price next year is.
$P_1 = (\frac{P}{E})$ratio × EPS= 16 x (\$7.55) = \$120.77.

4.12.8 As mentioned in this chapter, among the FCFF, FCFE or dividends models, the FCFE method is preferred when the dividend policy of the firm is not stable or when an investor owns a controlling interest in the firm. The dividend distribution is influenced by shareholder activism, taxation policy, and competitive capacity limitations.

4.12.9 It is most important to use multistage dividend discount models when valuing companies with initial high-growth rates. These companies are in the early phases of their life cycles, resulting in relatively rapid growth and low dividends or no dividends at all. Later, when the firms mature, growth rates decline to normal in the long run. Therefore, multistage models are used.

4.12.10 Suppose the intrinsic value of the stock is equal to its price. In that case, the market capitalization rate is similar to the expected rate of return if the stock is underpriced (i.e., intrinsic value < price). The investor's anticipated rate of return is greater than the market capitalization rate.

4.12.11 a. What is the intrinsic value of a share of stock?
calculate $k = r_f + \beta\ [E\ (r_m) - r_f]$ = 0.08 + 1.2 × (0.15 − 0.08) = 0.164 or 16.4%.
Dividends = D = 0.4 × \$10 =\$4.0
Growth rate = g = retention ratio × ROE = (1 – 0.4) x 0.2 = 1.2
Price = P = D(1 + g) /kg = 1.2 x \$4.0/(0.164–0.08) = 101.82.

b. The dividends next year
$D_1 = D_0(1 + g)$ = \$4*(1.12)= \$4.48
P_1 = V1 = V0(1 + g) = \$101.82 × 1.12 = \$114.04.
E (r) = (D1 + $P_1 - P_0)/P_0$ = (4.48 + 114.04–100)/100 = 0.1852 or 18.52%

References

Bruner, R. F., K. M. Eades, R. S. Harris, and R. C. Higgins. 1998. Best practices in estimating the cost of capital: Survey and synthesis. Financial Practice and Education 14–28.

Damodaran, A. 1994. *Damodaran on valuation*. New York: John Wiley & Sons.

Dimson, E. 1979. Risk measurement when shares are subject to infrequent trading. *Journal of Financial Economics.* 7 (2): 197–226.

Elton, E., M.J. Gruber, and J. Mei. 1994. Cost of capital using arbitrage pricing theory: A case study of nine New York utilities. *Financial Markets, Institutions and Instruments* 3: 46–73.

Estep, T. 1987. Security analysis and stock selection: Turning financial information into return forecasts. *Financial Analysts Journal* 43: 34–43.

Fuller, R.J., and C. Hsia. 1984. A simplified common stock valuation model. *Financial Analysts Journal* 40: 49–56.

Godfrey, S., and R. Espinosa. 1996. A practical approach to calculating the cost of equity for investments in emerging markets. *Journal of Applied Corporate Finance* 9 (3): 80–81.

Hamada, R.S. 1972. The effect of the firm's capital structure on the systematic risk of common stocks. *Journal of Finance* 27: 435–452.

Haugen, R.A. 1990. *Modern investment theory*. Englewood Cliffs, NJ: Prentice-Hall.

Kaplan, R.S., and R. Roll. 1972. Investor evaluation of accounting information: Some empirical evidence. *Journal of Business* 45: 225–257.

McConnell, J.J., and C.J. Muscarella. 1985. Corporate capital expenditure decisions and the market value of the firm. *Journal of Financial Economics* 14: 399–422.

Modigliani, F., and M. Miller. 1958. The cost of capital, corporation finance and the theory of investment. *American Economic Review* 48: 261–297.

Rosenberg, B., and V. Marathe. 1979. Tests of capital asset pricing hypotheses. Research in Finance 1:115–124. Siegel, J. Stocks for the Very Long Run: The Definitive Guide to Investment Strategies. New York, McGraw-Hill, 2007.

Part III

Losses, Recovery and Prevention

5 Risk and Monetary Impact

Learning objectives:

- Provide practical definitions of monetary loss
- Distinguish among returns and losses
- Appraise students with loss Measurement Tools
- Enumerate commonly used loss ratios for evaluating desk trades
- Enumerate and exemplify backtesting methods and applications
- Enumerate commonly used loss ratios for evaluating desk trades
- Identify Loss Limits and their application
- Thrust upon volatility adjusted stock value
- Infer the taxation aspects
- Emphasize the psychology of trading
- Explain the process of recovery from memory and unforeseen losses

5.1 Monetary Loss

Losses are measured after the lapse of an event (Shefrin and Meir 1985). The trading desk is risky for many untrained professionals. Loss is inevitable in life. In day-to-day life, one may lose money when he misses a flight, misses a deadline to pay monthly bills, leaves off a mobile phone or new pair of spectacles in a public toilet, etc. When one visits a doctor or buys medicines, he spends on laboratory tests that may end in fruitless diagnosis or endless treatment. Such expenses are also termed non-ordinary losses in life. Losses happen as accidents caused by serial lapses or more than "one miss" or joint misses. The disciplined trader is aware of the slightest chance of losing hard-earned profits. It takes days to recover, and it would largely depend on the spell of markets. It is advised not to lose more than a small one digit percentage of the money.

D. Bag, *Valuation and Volatility*,
https://doi.org/10.1007/978-981-16-1135-3_5

5.2 Return

The real test of a strategy is known only when it earns excess returns after making up for the trading or transactions costs. There are ways to measure returns for assets, absolute gains, relative gains and risk-adjusted gains, etc., absolute return is the return over a certain period.

5.2.1 Absolute Return

Absolute return is the simple difference between closing prices at two different points of time.

This measure looks at the appreciation or depreciation, expressed as a percentage. The absolute measures of return will vary from relative measurements because it deals with the return of one asset alone. It cannot compare the return values to other assets or other control situations.

5.2.2 Relative Return

Relative return is the simple comparison of absolute return and market return. The benchmark need not be the displayed exchange index or the displayed sectoral indices always. It can be a reference which the trader selects based on its consistency and relies on the benchmark for its trend.

Relative return is important because it is a way to measure the performance of actively managed funds, which should earn a return higher than the market. The relative return is a way to gauge a fund manager's performance.

5.2.3 Weighted Return

The weighted return looks at the proportion of money or the time duration for which the money remains in the account, including the timing and volume of cash inflows and cash outflows. The events of cash inflows are the first purchase, dividend inflows, cash withdrawals, re-deposits, terminal cash outflow, etc. The inflows and outflows are weighed by their periods of lock-in within the account. Money-weighted return is a scale of the relative return of an investment.

For example, it is given as $\sum_0^T \frac{CF_t}{1+irr^t} = 0$, where *IRR is* the rate of return.

5.2.4 Expected Return

The expected return is the mid outcome between a good gain and a bad loss. This is calculated using the formula.

$$E(r) = (\text{ProbabilityofGain})\,x(\text{NetGain}) + (\text{ProbabilityofLoss})x(\text{NetLoss}) \quad (5.1)$$

where the probability of loss (or gain) is the ratio between the count of loss (gain) events and the total number of events (both loss and fain) inferred from historical data.

Past performance is not indicative of future results. Any strategy that worked well in the past is likely to work well in the future, and conversely, any strategy that performed poorly in the past is likely to perform poorly in the future.

5.3 Measurement Tools and Backtesting

The tools for loss measurement are not too many. A few of the handy tools in the field include loss numbers of net loss, percentage loss, etc. There are indicators adjusted for the associated volatility. The list may consist of maximum upside and downside limits, win–loss ratio or the spectrum of risk-adjusted returns, etc. Backtesting is applying a trading strategy using historical data to check for the accuracy of the strategy. The test would predict actual results. Backtests compare trade results with model generated risk measures to assess the accuracy of predictions. The backtest models are relooked periodically, monthly or quarterly, to verify their accuracy. It is counted as the number of accessions, or departures, from the pre-specified confidence bands. The accuracy is examined within a pre-specified confidence band for value at risk (VaR) models that could compare financials such as daily P&L against the predicted Varro.

5.3.1 Risk–Reward

The risk–reward ratio is the prospective reward a position can realize. It is used to compare the expected returns against an amount of risk. It is the ratio of the amount expected to lose and the profit made at the sale off.

For example, a ratio of 1:7 implies that for the sake of earning $7, one risks $1. Similarly, a ratio of 1:3 means that one is prepared to lose $1 to earn $3. A good risk–reward ratio is higher than 1:3.

Suppose the trader has purchased 100 shares at $20 and placed a stop-loss order at $15 to make losses not to exceed $500. If the price rises to $30, the trader is determined to lose $5 per share and make a return of $10 per share. The risk–reward ratio is given as 1:2 (5:10) on the trade. If the risk–reward ratio is 1:5 (5:25), he will set the stop-loss order at $18 instead of $15.

5.3.2 Win–Loss Ratio

It is the ratio of the count of winning trades to the count of losing trades. When there are ten trades and six wins and four losses, the win–loss ratio is 6:4 or 3:2. The win–loss ratio does not mean that trader is "unsuccessful". Similarly, winning trades does not imply "much success". The win limits and loss limits are known to the trader himself.

5.3.3 Days' Average

It is the ratio of the number of days in which the portfolio beats an index by the total number of days over a period. The higher the days average, the better. The days' average of 0% means it never beats the benchmark. A day's average of 50% is used as a minimum threshold for measuring success. It focuses only on returns and does not consider the magnitude of losses faced by the portfolio.

5.3.4 Profit–Loss Ratio

It is a ratio of the average profit from winning trades and the corresponding average losses on losing trades in a given time. The profit–loss ratio does not consider the number of wins or losses. For example, a trader with a win average of $800 and a loss average of $400 over a period will have a profit–loss ratio of 2:1. A profit of $10,000 in earnings for a month and $6,000 in losses will give a ratio of 1.6. Alternatively, the loss recovery rate in a trade is the ratio between tax benefit claimed and recovered to total drawdown. The profit–loss ratio is different from the average profit per trade, which is the average amount to win or lose for each trade.

5.3.5 The Sharpe Ratio

Sharpe ratio is the measure of the risk-adjusted return of a portfolio. It is the division of the return (annualized) over the risk-free rate (R_f) by the volatility.

$$S_R = \left(\frac{R_P - R_F}{\sigma}\right) \tag{5.2}$$

where S_R is the Sharpe ratio.
Where

- R_p is the portfolio return,
- R_F is the risk-free rate of return. The 10-year Govt. bonds rate could be the risk-free rate.
- σ is the standard deviation of the portfolio.

The Sharpe ratio does not differentiate between upside and downside risks. An asset with a negative correlation improves the Sharpe ratio. A Sharpe ratio higher than 3.0 is good.

If the return is 25%, the standard deviation is 10%, and the risk-free rate is 5%, the Sharpe ratio is given as.

$= \frac{25-5}{10} = 2.0$. The limitation of this ratio is its inability to predict future losses.

5.3.6 Sortino Ratio

It is defined as the risk-adjusted return of a portfolio (Vince 1990). This ratio penalizes returns falling below a user-specified target or required rate of return,

When Sortino Ratio is higher, the probability of a large loss is lower.

$$Sortino = \left(\frac{R_P - R_F}{\sigma_D} \right) \tag{5.3}$$

where

R_p is the portfolio return,
R_F is the risk-free rate,
σ_D is the standard deviation of negative asset returns (downside deviation).

Sortino ratio includes the downside deviations (standard deviation of negative returns) instead of the standard deviations of the full sample. A Sortino ratio of value higher than 2.0 is good.

5.3.7 Calmar Ratio

The Calmar ratio is defined as the excess return divided by the maximum drawdown.

$$Calmar = \left(\frac{\mu - r_F}{\max(\text{Drawdown})} \right) \tag{5.4}$$

The maximum drawdown (MDD) is the maximum observed loss from a peak to a trough of a stocks' price that occurs before a new peak. A simpler situation is when the expected return is $x\%$, the corresponding expected drawdown is also $x\%$. Calmar ratio higher than 0.50 is good.

5.3.8 Sterling Ratio

The sterling ratio is the excess return divided by the average MDD (maximum drawdown) in a year. The average maximum drawdown is calculated over three years. A simple upfront deduction of 10% is applied. It considers the return (annualized) over the last 3 years by dividing it into split windows of three independent 12-month periods. The maximum drawdowns are independently calculated for each window to report the average.

$$\text{Sterling ratio} = \left(\frac{\text{Annualized Return}(\%)}{\frac{\text{Drawdown}_1 + \text{Drawdown}_2 + \text{Drawdown}_3}{3} - 10\%} \right) \tag{5.5}$$

where, $Drawdown_1, Drawdown_2$, and $Drawdown_3$ are the drawdowns for the last three years.

A sterling ratio higher than 1.0 is preferred.

5.3.9 Omega

Ω(r) is the ratio of the cumulative probabilities above and below the specified threshold. It is the ratio of the weighted probability of gains to losses for any return level r.

Omega is given as.

$$\omega(r) \frac{\int_{-a}^{x} [1 - F(x)] dx}{\int_{-a}^{x} F(x) dx} \tag{5.6}$$

where *F(x)* is the cumulative distribution of returns,

r is the threshold level of return.

At higher threshold levels of r, the Omega is lower. It does not assume the underlying distribution of returns. The ranking of portfolios using Omega is different from that of the Sharpe ratio.

5.3.10 K-Ratio

A value-added monthly index (VAMI) is created from a hypothetical $1000 investment. The monthly performance of a VAMI is reported over at least 30 months. K-ratio is the product of $t_{\text{calculated}}$ $\left(\frac{\mu}{\text{Std.Error}}\right)$ and the square root of the number of observations ($\sqrt{N}$) in the sample.

The β_{VAMI} are estimated from the trend line, which is a regression equation given below,

$$Log(R)_{VAMI} = \alpha + \beta_T Log(VAMI)_T + \varepsilon \quad (5.7)$$

where Log (R_{VAMI}) is the log of cumulative return on VAMI.

β_{VAMI} is the Slope of Log (VAMI) Regression Line,

σ_{VAMI} is the Standard Error of β_{VAMI}.

$\sqrt{N}\ ^2$ is the Square Root of the Number of Observations in the sample.

The $t_{calculated}$ equals the ratio $(\frac{\beta_T}{\sigma_{VAMI}})$ which follows the student's t-distribution. Higher is better.

5.3.11 Treynor

Treynor ratio, also known as the reward-to-volatility ratio, is a performance metric for determining how much excess return was generated for each unit of risk taken on by a portfolio.

$$Treyner\,Ratio = \left(\frac{R_P - R_F}{\beta}\right) \quad (5.8)$$

where

R_P is the return of the investment,

R_F is the risk-free rate of return,

β is the beta of the portfolio.

A Higher Treynor ratio is better.

5.3.12 Jensen's Alpha

The Jensen's alpha is the classic CAPM-α, a positive number that ensures better return compared to benchmarks in the market. It is given as

$$\text{Alpha} = \alpha = R_i - R_F - \beta(R_M - R_F) \quad (5.9)$$

where

R_i is the realized return of the portfolio,

R_M is the realized return of the market portfolio,

R_F is the 10-year Govt. Bond yield,

β is the beta of the portfolio.

5.3.13 Rachev Ratio

It is the Expected Tail Return per unit expected Tail Loss. It is based on the value at risk (VaR). The expected Tail Return (ETR) is the average of the right 5% of the distribution of returns. The expected Tail Loss (ETL) is the average left 5% of the distribution of returns during the period. Conditional VaR is the loss that could happen conditioned on the loss amount exceeding the VaR limit of the loss distribution. Rachev ratio is the ETL of the opposite of excess return at a given confidence level, divided by the ETL of the excess return at another confidence level.

$$Rachev\ ratio(\rho) = \frac{cvaR, 5(R - Rf)}{cvaR, 5(Rf - R)} \tag{5.10}$$

where $cvaR_5$ is the conditional value at risk at 5% probability.

5.4 Loss and Margin

In margin trading, the broker funds the balance. The exposure is arrived at multiples of the initial margin deposited. The margin is about 20–25% in stocks, with the balance being funded by the broker. In the case of futures trading, the margin is around 15–20% of the value of the stock. If the price movement follows a downward trend, the call is to deposit mark to market (MTM) margins to make up for the negative price movement. In the case of margin trading, there is no concept of MTM margins. If the client is not able to meet the MTM margins, the broker is authorized to close out futures position and debit the losses to the client account. The list of stocks included in margin trading is different from the list included in futures trading. Margin trading has the benefit of carrying forward the facility of existing positions.

In contrast to margin trading, direct trading in futures is better because it entails a higher cost in rollover costs. The margin at portfolio level allows setting aside the lender's exposure by consolidating or netting positions to display the lower risk than actual. The result is lower margin requirements for hedged positions. For example, if a position in the portfolio is net positive, it could offset the liability of a losing position in the same account. This would reduce the overall margin requirement that is necessary for holding a losing derivatives position. The minimum margin is maintained in the account of a naked option writer at 100%. Margin is not required for covered writing option positions. The normal margin for underlying uncovered stocks is 25%.

5.4.1 Loss Limits

It is not possible to avoid trade losses completely and permanently (Balsara 1992). If one manages to save money in a down market, he is far ahead of many others in the race. After a fixed portion of the principal traded amount is lost, the trader stops trading and exits the loss-making position. In the case of pre-market instruments of overvalued IPOs bought at a premium, the prices on a listing day may turn lower and live losses. In case of overvalued IPOs bought at a premium, the prices on listing day turn lower and are related to losses. The institutional traders on D day are the ones whose horizon is lower than that of retail IPO subscribers. First-time players have fund issues on IPO exposures and may decide to exit on the listing Day. FIs and FPIs choose to dispose of portions of their pre-market holding on a listing day, and there are no physical penalties on them for off-loading their holding limits.

5.4.2 Maximum Drawdown (MDD)

Maximum drawdown is an indicator of downside risk over a holding period. A maximum drawdown (MDD) is the maximum observed loss from a peak to a trough of a stock price that occurs prior to a new peak. If an investment has never lost a penny, the maximum drawdown would be zero. The worst possible maximum drawdown would be 100%. The losses from the investment are small when the maximum drawdown is low. The umbrella Loss Limit is an overarching limit across combinations of multiple positions. It is applied on top of stop-loss limits.

5.4.3 Zero Variance Portfolios

The variance of an individual stock is never zero. A trader's utopia is to have a zero variance portfolio. They could be close to the minimum variance portfolio or have low correlations with each other. Zero variance portfolios are different from zero beta portfolios. A zero beta portfolio aims to include negative beta stocks in the bundle of positive beta stocks to make the final beta close to zero. The unsystematic risk can be minimized with substantial diversification, where the effective beta is much lower than the initial.

5.4.4 Risk-Adjusted Value

The adjusted risk measures are more sophisticated than simple algebraic manipulations that include (1) expected shortfall, (2) Mean shortfall, and (3) Mean exceedance, respectively. The "mean shortfall" covers the left tail, which is the mean of returns below the p-th percentile, is the complement of the expected shortfall, known as conditional value at risk. The value at risk measures the p-th percentile of the returns distribution is; that is, it ignores returns below the p-th percentile. The mean shortfall measures left-tail, which is most relevant to be used. It reduces the likelihood of extreme returns (and the volatility of volatility). In particular, the lower probability of very negative returns (left-tail events) is commonly used for investors.

5.4.5 Value at Risk (VaR)

The value at risk (VaR), a probability-based numeric measure that assesses the probability that the losses to be made, would not exceed a cutoff (threshold) loss value. VaR is observed from the loss distribution. For one day, when the VaR is 1%, there is a 1% chance of loss to exceed $10,000, a loss limit. This means that the 99% confidence that his maximum daily loss will never go beyond $10,000. Therefore, vaR gives a snapshot view of the given position, which does not remain constant. VaR could be denoted as absolute vaR or relative VaR.

The absolute VaR is arrived from the distribution of historical returns known as the right tail of the loss distribution. The 99th percentile, a price equivalent to a 10-day minimum "holding period", is used. The longest observation period (sample period) for historical data for a length of one year is used. The various forms of relative VaR include (1) incremental VaR (IVaR), (2) marginal VaR (mVaR), (3), conditional VaR (CVaR), etc. The incremental VaR denotes additional risk (vaR), a new position or stock that adds to a portfolio. MVaR measures the change in portfolio VaR for a given change in the portfolio composition. mVaRs are added together to report the share of each asset to the overall VaR. Conditional VaR is the loss that could happen conditioned on the loss amount exceeding the VaR limit of the loss distribution. Conditional VaR is akin to conditional probability, where the average loss is conditional on exceeding a given VaR limit. The vaR limit is the confidence limit of the distribution. The limitation of vaR includes assessing potential loss that represents the lowest amount of risk in a range of outcomes. For example, a VaR of 95% with 20% loss represents an expectation of losing at least 20% one of every 20 days on average means the same with a given loss of 50%.

5.4.6 Liquidity Value at Risk

Liquidity is the ability to escape quickly from the stock. When volatility rises, it causes liquidity risk to many stocks due to lower trading volumes. The base vaR does not account for liquidity (Coughlan 2004). The holding period of the exposure increases due to such risks. The component of liquidity vaR is added to the base vaR to represent the total risks. As this time horizon is increased (due to the illiquidity of the portfolio), the reported VaR also increases to reflect the higher risk. A liquidity margin equal to a 10-day holding period is used.

5.4.7 Fixing Limits

Limits are proportional to the expected margin or losses due to exposures. The individual limits are derived from the limits imposed by the member brokers, which in turn flows from the exchange (Chande 2001). For example, the margin over a day is the algebraic sum of (1) VaR margin, (2) Extreme Loss Margin and, (3) mark to market margin, respectively. When the margin is applied on the cumulative open position of a trading member, the cumulative open position is the total of net positions attributed to individual trading clients of the member. The cumulative open positions do not consider netting or set of positions against positive or negative exposures.

$$\mathrm{MTM}_{\mathrm{Loss}} = (\mathrm{Qty}_{\mathrm{Buy}} \times \text{closeprice BuyValue}) - (\text{SaleValue - (QtySale} \times \text{closeprice})) \tag{5.11}$$

The exchange links the scrip VaR with the index vaR daily. For low volatile scrips, the minimum is 5%. The daily VaR is a minimum of 7.5% or multiplying by 3.5, the average daily volatility. For high volatile scrips, the daily VaR is the higher between scrip VaR (3.5 sigma) or 3.0 times the VaR of the index, which is multiplied post determination by a fixed value ($=\sqrt{3}$).

The effective cost of funds (leverage) can relate to the VaR limits on trades for levered positions with borrowed funds. The daily VaR limit is 3.5 times the volatility for a minimum of 7.5% for each scrip. For example, if the leverage is five times on a trade (1:5), a 10% negative movement in the price can result in a 50% erosion of the margin money. The spot price that is displayed is higher than a traders' reserve price level.

5.5 Path to Recovery

Each one of us can handle infrequent occasional minor losses. When one is yet to recover from one loss and the other loss piles on, it requires coping skills for the individual. No one can ever buy insurance for all losses in life. This shows that as losses get larger, the return necessary to recover from breaking even increases at a much faster rate. For example, in terms of expected percent gains, a mere 10% loss would expect at least a percent net gain of 11% to assume the initial state. Taking time off from the trading desk for a few days could make a difference to impulses. The impact of staying away from the desk for a few months is only to forgive oneself. Once they come back to the trading desk, they need to trade small. One needs to average out his funds in smaller increments all along. For example, this means to deploy a total sum of USD 1000; one has to pump in installments of USD 100 each, ten times.

When a trader takes on a bad trade, he loses self-confidence. The plan to pass through tough times could start with the creation of a cash buffer. The simplest task is to sell off non-performing stocks right away. It is prudent to rebalance the bundle by selling off shares that had underperformed. It is impractical to divert funds from long-term needs such as housing construction, property development, paying for higher education, un-matured insurance, etc. Financial assets such as fixed deposits, Mutual Market Mutual funds (MMMF), gold deposits, ETFs, commodities, forex, etc., Money market funds (MMMF) are considered less risky. MMMFs provide higher returns and are also preferred by institutions, etc., and can be liquidated as they are more volatile than the stock. Gold is most convenient to dispose of by individuals. Regulators may not permit direct trading in forex by individuals. It is prudent to desist to borrow from informal sources. Institutions could also resort to the disposal of physical assets or financial assets. They are matching principles that apply to recovering from medium-term losses with short-term gains because short-term gains can be redeployed further. Institutions have a longer horizon, and they can plan to make good the loss over a longer time. Lastly, Institutions can make provisions for the losses caused or write off the losses from their books. Institutions can take advantage of commercial property because they can recourse to sale-leaseback. Institutions can claim the losses made on the part of their employees from insurance companies.

Table 5.1 shows the Carry forward Monetary Losses.

Large telecom operators may never come back at lower levels of ARPU. When investors hang on to stocks with large unrealized capital losses, “dead money” never recovers.

One is not prepared to face failure and the stigma of failure. The ability to bounce back is a measure of emotional intelligence. Many traders will never come back to their desks. Post recovery therapy will boost morale. One can pick up good trading books to learn theory, join classes and courses to build on learning, education or attend coaching sessions. This will increase trading confidence and experience. Attempts to revenge trading will compound initial losses.

There could be a plan to turn around, which might take more than a year or so to get back those losses. The traders use technology to reduce the doggery associated with trading and keep track of trades that have already been made. When back at the trading desk, one needs to trade small. Capital losses cannot be made good by revenue contribution. The monthly exposure to stocks is only a maximum of 20% of the rental earning from your property assets, as a personal limit.

There are hard ways to make good the loss for an ordinary individual partially. However, when one is down with losses, the rental earnings help recover over a horizon of one year or so only when the terms of the rent /lease agreement are revised. The existing terms are stretched a bit with the tenant. The tax benefit to taxpayers on borrowed loans from property does exist. When the taxpayer possesses rental properties, tax laws permit standard losses on house repair loss @30% of the annual rental income. A family man has to discover ways to reduce his monthly family expenses by cutting down a few components of his routine lifestyle to manage himself over the days. To cut down on supplementary expenses, one can

Table 5.1 Carry forward monetary losses

Description	Transaction type	Buy date	Buy Qty	Cost per share	Purchase commission	Total cost	Sell date
Test 1	LIFO	8/11/2014	110	$12.00	$5.99	$1,325.99	9/13/2015
Test 2	Average Lot	7/10/2015	50	$17.00	$5.99	$855.99	6/14/2015

Description	Sell date	Sale Qty	Proceeds per share	Sale commission	Total proceeds	Gain/loss ($)	Gain/loss (%)	Term
Test 1	9/13/2015	110	$13.57	$5.99	$1,486.71	$160.72	12.12%	Long term
Test 2	6/14/2015	50	$15.21	$5.99	$754.51	($101.48)	–11.86%	Short term

surrender subscription fee expenses on products such as golf memberships, club memberships, surrendering platinum credit cards or parking licenses, monthly magazines, cut down on annual fixed commitments and lifestyle weekend expenses, etc. It is advisable to hold time deposits in fixed deposit accounts or debt exposures so that a small portion of the cumulative interest earned or accrued from matured products can be adjusted against trading losses. This means one can avail low-cost top-up loans on existing active term loan accounts to adjust for the losses in trading.

5.6 Reducing Trading Costs

Trading costs are an essential part of the overall costs that could make the difference to net returns. The trading costs impose a drag on portfolio returns. Full-Service Brokers provide a gamut of services to their clients and charge higher brokerage fees. Many of the funds in the market may underperform due to similar costs. Trading costs include 1. brokerage costs, 2. costs of the bid-ask spread, 3. price-inventory impact due to volume changes of executed trades, and 4. opportunity cost of locked funds. The burden of trading costs is more for short-term holding than long-term holding. Since the aim is to reduce trading costs, discount brokers or budget brokers can offer trading platforms at affordable rates.

For example, Zerodha, 5Paisa, Angel Broking, Trade Smart Online may or may not exist for long in the marketplace.

The choice of passive investing is available to the trader to lower trading costs. To invest in mutual funds, if a Full-service brokerage service provider owns the fund house, the transaction charges for customers could be lower. Direct trading is cheaper than hiring advisory services. A single serial order of a bundle of stocks entered together is cheaper than individual stocks against multiple orders. The opportunity costs of parked funds increase with margin trades. Opportunity costs include liquidity costs that may arise to fulfill personal consumption needs when the funds are blocked in trades positions. With margin investing, the leverage costs are added, and so are the penalties for late fees added by the broker for missing deadlines on margin calls. One loses more in levered trading in case of Drawdowns. Informal sources of finance are costlier. Alternatively, the trader loses a golden opportunity to gain from rising prices displayed live at the exchange without margin.

5.7 Memory and Recovery

People never forget their past losses easily (Thaler 1999). Telescoping is an approximation route to standby record-keeping of transactions. Telescoping is used in accounting practices to arrive at projected income or to project losses. The entries made in the journal can remind the trader of past failures and have adversely impacted him. It takes care of missing entries and does not use statistical methods to replace the missing entries values.

In many unmarked credit and debit entries, a peak (maximum) method is applied that considers the maximum amount. When there is an unexplained income of an in the applicant's books that appears in the first part of a year and a corresponding unexplained purchase of a similar amount that appears in the latter part of the year, it is considered that the unexplained transaction has been made out of the unexplained income. The benefit of credit entries is given to the trader because of his loss of memory. For example, it is usual to consider the highest credit entry into the small retailers business in loan appraisals to arrive at his average income.

Table 5.2 shows the calculating intra-day loss due to trading in two different stocks.

5.8 Telescoping Bias

There exists a memory bias in remembering losses or events, which is denoted as the telescoping bias. People are more sensitive to losses than to gains. Individuals exhibit narrow framing, i.e., narrowly defined gains and losses than proprietors. Investors become more loss-averse if their previous investment had made losses. Institutional managers are more risk-averse than retail investors since they deal with client accounts and would face the risk of reputation loss. In cognitive psychology, the telescoping effect (or telescoping bias) refers to the displacement of an event whereby people perceive recent events as being more remote than they are.

5.8.1 Backward Telescoping

Backward telescoping refers to how long a person can remember events back in time and how he perceives the lapse of time. Neter and Waksberg (1964) have highlighted the phenomenon of bounded recall to ask participants about events, and the participants are reminded of past events and their related occurrences. The telescoping errors are minimized when an individual has more than one reference point to date major events. The major reference events are marriage, death or family life events. Similarly, they perceive distant events as being more recent than they are. The former is known as backward telescoping or time expansion and the latter is known as forward telescoping.

5.8.2 Forward Telescoping

People have a systematic tendency to recall that distant past events had occurred recently (forward telescoping). Recent events occurred farther back in time

Table 5.2 Intra-day profit calculation from two stocks

Sr. No	Name of the script	Purchase rate	Purchase Qty.	Gross purchase value	Brokerage	Service tax on brokerage	Purchase value	Sale Rate	Sale Qty.	Gross scale value	Brokerage & Service Tax	Sale value	Gross Profit/ (Loss)	STT	Stamp duty charges	Transaction charges	Service Tax 2	Net Profit/ (Loss)
					0.03%	10.30%					0.03%			0.025%	0.010%	0.008%	0.007%	
1	ITC	100.00	100.00	10,000.00	3.00	0.31	10,003.31	110.00	100.00	11,000.00	3.30	10,996.36	993.05	2.75	1.10	0.88	0.77	987.55
2	SBI	120.00	100.00	12,000.00	3.60	0.37	12,003.97	110.00	100.00	11,000.00	3.30	10,996.36	1,007.61	2.75	1.10	0.88	0.77	(1,013.11)

(backward telescoping) than the actual, including events beyond the period, either events that are too recent for the target time event (backward telescoping) or events that are too old for the target period (forward telescoping). Memorable events are recalled as occurring recently. People underestimate memory loss over long periods, and target events are moved closer to the current time. People date events nearer to a reference target period, such as a year or a vacation or a big event in life, similar to linking target events to other events in life.

People tend to underestimate the length of long time intervals and overestimate the length of short time intervals. Older children have a tendency to telescope earlier memories and a weaker tendency to telescope recent memories than younger children. Children do recollect but cannot easily give a chronological account of past events. Older perceive longer events as more recent. Many older adults claim that time speeds up as they get older, explained by forward telescoping. Since forward telescoping leads people to underestimate the amount of time that has occurred since an event, this is the reason why people perceive time as moving faster as they age. The other examples include forward telescoping in reported age of first use of smoking, alcohol, drug use, etc. The other uses in marketing, when respondents often inaccurately describe the past purchase of products is known as forward telescoping.

5.8.3 Psychology of Loss Making

Nobody wants losses it to happen over and over again. The psychological impact of the loss of physical gold or concrete property is more than financial assets. If the first loss is higher than the second loss, then the psychological impact is lower. Else, if the first loss is of a lower amount than the second loss amount, then the impact is much higher. The disposition refers to investors' reluctance to dispose of assets that have lost value and the preference to sell assets that have made gains (Shefrin and Statman 1985). This is elaborated by loss aversion, guilt, regret, etc. Investors are susceptible to bias when they view recent gains as disposable gains deployed to purchase high-risk investments.

One mistake alone does not cause an accident; instead, a series of correlated or serial mistakes can cause loss. Investors often show irrational behavior. They tend to be influenced by psychological factors, such as overconfidence, greed, cognitive dissonance, representativeness bias, self-attribution bias, etc. Individuals do have self-attribution bias, which is desirable to be successful in trading. Beginners may trade actively to cultivate confidence. This makes them over trade and herding. There are both personal and situational reasons for making impulsive trades.

Orthodox traders fall into cognitive bias or hindsight bias leading to past beliefs. Many old convictions on pharma and biotech stocks do cling to the notion of the glorifying future that exclusively depends upon new launches of new drugs against new ailments. Hindsight Bias is the unknown that seems known. When the event occurred in reality, the event is considered predictable.

5.8.4 Stop Repeating Losses

Human beings memorize successes and seem to forget past failures. One should not Never undertake trades where one is not confident of gaining; needs to weigh the arrival of new information appropriately by looking for evidence that contradicts the existing position. There is no shortcut or a known path to guaranteed recovery of the full losse. However, one must attempt partial recovery of the losses. Prevention of losses is easier than post facto recovery of losses. The discipline of record-keeping of date and time stamp of trades can establish the proportion of impulsive trades undertaken on occasional Fridays. Post-trade analysis can be reviewed across executed trades to give insight to portfolio managers to control them.

Corrections in the market can lead to panic, overselling harms the short-term investors. The average market correction is short-lived and lasts between three and four months. Leveraged traders can turn into a prolonged decline. There may not exist insurance for covering direct trading losses. Although we worry about making losses, we need not panic, deter, and run away from markets. Short-term tax benefits on capital losses always exist that can come to pour rescue and continue to support us through our bad times.

Large institutions can partly recover the expected costs of trading losses by incorporating in the marketing costs, ex-ante, or acquisitions costs of new clients or variable pricing terms for segments of clients. Individual traders have to resort to informal sources, private sources, compromise on consumption, lifestyle, and exorbitant borrowing rates from informal sources or liquidating non-financial assets to make good money. In no case, rational investors should adopt panic selling of non-financial assets to recover from the loss. As mentioned before, the psychological preparedness for the emotional tragedy of loss is far more important than the money loss.

5.9 Summary

This chapter provided insights on the terminal monetary impact on the investors. This chapter also described the end period's impact on investors' money. This represented the major tools and instruments of measurement. It explaineds Value at risk Limits. Volatility adjusted Valuation and the measurement of Loss of Value to Stakeholders. This highlighted win–loss ratios, drawdowns, Sharpe, Calmar, sorting, omega measures. There are various ways in which an investor can improve his risk-adjusted return. The emphasis was on the horizon, adjusted measures, and methods to translate to money impact. It highlighted that futures trading in stocks meets the criteria of profitability, quality, liquidity, etc. It narrated the loss-making process, the psychological impact of losses, the trading psyche, and the recovery plan. It gave examples of households and institutions.

5.10 Questions

5.10.1 How do you apply the 6% loss limit to a portfolio?

5.10.2 When the number of trades is different, how do you apply the 2% rule?

5.10.3 How to construct a zero variance portfolio?

5.10.4 Calculate the vaR limit from below?
Suppose an asset has a 3% 1 month VaR of 2%, representing a 3% chance of the asset declining in value by 2% during the 1 month time frame. The conversion of the 3% chance to a daily ratio places the odds of a 2% loss at 1 day per month.

5.10.5 Jensen's Alpha is calculated as the difference between the portfolio's return and the actual return. Calculate the portfolio's Alpha from the Data Given Below?
ER (A) = 0.05 + 0.95*(0.1 – 0.05) = 0.0975 or 9.75%
ER (B) = 0.05 + 1.05*(0.1 – 0.05) = 0.1030 or 10.30% return.
ER (C) = 0.05 + 1.1* *(0 .1 – 0.05) = 0.1050 or 10.50% return.

5.10.6 Calculate the VaR margin amount from the data given below on a purchase value of 1 80,000?
For ABC Ltd, the calculated vaR rate was 13% based on volatility. When the standard deviation of the logarithm of daily returns of the security is 3.1%. 1.5 times standard deviation would be $1.5 \times 3.1 = 4.65$.

5.10.7 8 in a drawdown, how much raise is necessary? If the capital loss is 30%, then the recap is 50%. If the maximum exposure is only 5%, you lose only 1% of the total value, which would cut the overexpo-sure to zero?8 in a drawdown, how much raise is necessary? If the capital loss is 30%, then the recap is 50%. If the maximum exposure is only 5%, you lose only 1% of the total value, which would cut the overexpo-sure to zero?

5.10.8 Suppose one sells short 500 shares of a stock, currently sell-ing for $40 per share, and trades with a broker for $15,000 to estab-lish his your margin account.

a. If you earn no interest on the funds in your margin account, what will be the rate of return after 1 year if the stock is selling at: (i) $44; (ii) $40. There are no dividends.
b. If the maintenance margin is 25%, how high can the stock's price rise before the margin call?
c. Now assume that the stock has paid a year-end dividend of $1 per share. Re-calculate, (a) and (b) above?

5.10.9 Probabilities for three states of the economy and probabilities for the returns on a particular stock in each state are shown in the Table 3. What is the probability that the economy will be neutral *and* the stock will experience poor performance?

5.10.10 Which of the below description of the goal of a delta-neutral portfo-lio is appropriate?

Table 5.3 State and Probabilities of Gains

State of economy	Probability of state	Performance	Probability of stock performance
Good	0.3	Good	0.6
		Neutral	0.3
		Poor	0.1
Neutral	0.5	Good	0.4
		Neutral	0.3
		Poor	0.3
Poor	0.2	Good	0.2
		Neutral	0.3
		Poor	0.5
Total			1

A delta-neutral portfolio is denominated when:

i. A long position in one stock is added to a short position in call options to make the portfolio's value-neutral to the stock price changes.
ii. Long position in a stock is added to a short position in a call option to make the value of the portfolio neutral to changes in the value of the stock.
iii. Long position in a stock with a long position in call options so that the value of the portfolio does not change with changes in the value of the stock.

5.10.11 In the beginning of the year, Mr. A decided to withdraw $5,000 in savings out of the bank and put in bonds and stocks. After one year, the value of bonds and stocks stood at 2300 and 2500, respectively. Further, $100 was received as dividends and 300 as coupon payments. What is the return on his portfolio?

5.10.12 Mr. D's portfolio is worth $12,000 at the beginning of a month. After 10 days, he pumped in $0.800 to the portfolio. At the end of the 30 day month, if the value has risen to $13,980, what is the dollar-weighted gain?

5.10.13 When to use time-weighted and dollar-weighted rates of return?

5.10.14 An investor A is willing to purchase stock ABC for $9 or $12 by year-end with a 50% chance. The investor has the option to borrow 50% from the broker @9%. What is the expected yield if he buys 200 shares by borrowing 50%?

Solutions

5.10.1. Stop losses are undertaken due to a fall in price. The cash that lies in the trading account, cash equivalents, and the last of all open positions in the account. The daily total for today is compared with the total on the last trading day of the previous month to check if the ratio is smaller or higher than the 6%. This means a 6% monthly loss limit translates to three open positions of 2% losses each, or six open positions or 1% losses each, and so on.

5.10.2 The 2% single trade limit means that the trading positions are re-calibrated every month. In one month, when there is a loss, the smaller remaining total cash in the trading account in the following month will ensure that your trading positions are smaller.

5.10.3 A zero variance portfolio can be constructed when the correlation coefficient between two assets is close to –1.0. When the first stock rises by 1, the second stock fall by 1. The change is 0. Since the value does not change, the variance is 0. The gains and losses are perfectly canceled out to have a net change equal to zero.

5.10.5 Since the asset has a 3% 1 month VaR of 2%, that represents a 3% chance of the asset declining in value by 2% during the 1 month time frame, the conversion of the 3% chance to a daily ratio places the odds of a 2% Loss at 1 day per month.

5.10.6 Jensen's Alpha is Calculated as Follows:
ER (A) = 0.05 + 0.95*(0.1 – 0.05) = 0.0975 = 9.75%
ER (B) =0.05 + 1.05*(0.1 – 0.05) = 0.1030 = 10.30%
ER (C) = 0.05 + 1.1*(0.1 – 0.05) = 0.1050 = 10.50%
Alpha A = 12 – 9.75% = 2.25%
Alpha B = 15 – 10.30% = 4.70%
Alpha C = 10 – 10.50% = –0.50%

5.10.7 For ABC Ltd, the calculated VaR rate was 13% based on volatility. When the standard deviation of the logarithm of daily returns of the security is 3.1%, then1.5 times standard deviation would be $1.5 \times 3.1 = 4.65$. Since higher is 5% (than 4.65%), the vaR margin for Extreme Loss margin rate will be 5%. Therefore, the total margin on ABC will be 18% (13% VaR Margin + 5% Extreme Loss Margin). As such, the total margin payable (VaR margin + extreme loss margin) on the trade of INR 10 lakhs would be INR 1,80,000/-

5.10.8 In a drawdown, if the maximum level of exposure is only 5%, you lose only 1% of the total value. This will cut down the overexposure overexpo-sure is to zero?

5.10.9 a. There are no dividends.

i. rate of return = (–500 x $4)/$15,000 = –0.1333 = –13.33%

ii. rate of return = (–500 x $0)/$15,000 = 0%

b. Total assets in the margin account are $20,000 (from the sale of the stock) + $15,000 (the initial margin) = $35,000. Liabilities are 500P. A margin call will be issued solving for (1 + p)500 = (35,000, or P = $56 or higher.

c. With a $1 dividend, the short position pays the borrowed shares: ($1/share × 500 shares) = $500.

A margin call will be issued when P = $55.20 or higher.

i. rate of return =[(–500 X $4) – $500]/$15,000 = –0.1667 = –16.67%

ii. rate of return = [(–500 x $0) – $500]/$15,000 = –0.0333 = –3.33%

5.10.9

It is the joint probability that the economy is neutral and that the stock performance will be poor. It is the product p = (0.4)(0.5) = (0.2) or 20%.

5.10.11 A Delta-Neutral Portfolio is Denominated When;

i. A long position in one stock is added to a short position in call options to make the portfolio's value-neutral to the stock price changes.
ii. Long position in a stock is added to a short position in a call option to make the value of the portfolio neutral to changes in the value of the stock.
iii. Long position in a stock with a long position in call options so that the portfolio's value does not change with changes in the value of the stock.

5.10.12 Final Sum Received After One year = 2300 + 2500 + 100 + 300,

Gain = 5200–5000 = 200

Return = 200/5000 = 2%

5.10.13 Dollar-weighted returns do reflect cash inflows and outflows. Let's divide a month of 30 days into three parts of 10 days each and determine the 10-day return of *r*.

The matching of values are given as.

13,980 = 12,000 x $(1 + r)^3$ + 800 × $(1 + r)^2$, which translates to r = 3.05.

Hence, the annual return are $k = (1 + r)^3 - 1 = 9.44\%$

5.10.14 The time-weighted return is a better measure for situations with no control over the size or timing of cash flows, e.g., listed funds. The money-weighted return is used when the investor exercises control over the amount and timing of fund flows.

5.10.15 The expected annual return E (r) = 50% x $12 + 50% x $9 = $12.5.

If the investor borrows 50% from the broker @9%, the cost of borrowing = 50% × 9% = 4.5%. The expected yield = 12.5–4.5% = 8.0%.

References

Balsara, Nauzer. 1992. *Money Management Strategies for Futures Traders.* NY: John Wiley & Sons.

Chande, Tushar. 2001. *Beyond Technical Analysis: How to Develop and Implement a Winning Trading System*, 2nd ed. NY: John Wiley & Sons.

Coughlan, G., 2004. Corporate risk management in an IAS 39 Framework, JP Morgan,2004.

Shefrin, Hersh, and Meir Statman. 1985. The Disposition to Sell Winners Too Early and Ride Losers Too Long: Theory and Evidence. *The Journal of Finance.* 40 (3): 777–790.

Thaler, R.H. 1999. Mental accounting matters. *Journal of Behavioral Decision Making* 12: 183–206.

Vince, Ralph. 1990. Portfolio Management Formulas: Mathematical Trading Methods for the Futures, Options, and Stock Markets. John Wiley & Sons, 1990

References

Hedging

6

Learning objectives:

- Provide brief narration of hedging and related instruments
- Emphasize on moneyness of short and long hedges
- Combine put and call options and netting
- Learn devise escape and exit methods
- Design and evaluate the tests of the efficacy of Hedging
- Understand Qualitative Methods of Hedge Effectiveness
- Enumerate Volatility Strategies

6.1 Derivative

Derivatives offer a method of effective protection against risks due to variation in spot (hazir) prices. It is a financial instrument that is derived from the values of an underlying instrument. The underlying security is a financial asset or commodity traded in an exchange (Merton 1973). The trader would find it difficult to accurately relate to the underlying unless the underlying is thickly traded. Hedging could be bought the moment a stock is bought or sold. Hedging instruments are broadly classified as forward contracts and futures contracts (financial securities). The futures contract is standardized by the stock exchange under its general terms and transparently executed.

6.2 Single Stock Futures

Futures Contracts are sold for individual stocks for specific prices for future dates. A single stock future (SSF) is a futures contract of a specified stock for a given expiry date. SSF is an inexpensive method to buy a stock. It protects against volatility or declines in the price of the underlying stock.

D. Bag, *Valuation and Volatility*,
https://doi.org/10.1007/978-981-16-1135-3_6

Let's see an example, if one bought 250 shares of ABC stock at INR 2,284 per share for INR 571,000, a "long" on ABC stock. In order to hedge the position in the spot, a short position in the futures market is purchased at a price of INR 2,285.

The number of Lot size (N) = 250, Contract Value = 2285 × 250 = INR571,250.

The net payoffs against different price points of the futures are given in Table 6.1.

Even when the price is headed downwards, the position will not lose money. The hedging of single stock positions is easier.

When there is more than one stock, the Beta (sensitivity) of the portfolios is used. Table 6.2 shows the natures and influence of betas.

Table 6.3 shows a portfolio of INR 800 000/- is invested across the following stocks. The portfolio of INR 800 000/- is hedged against NIFTY futures.

The β_P of the portfolio is derived as the sum product of betas (βi) of each stock with its weight in the basket.

The hedge value is product of Portfolio Beta (β_P) and portfolio investment = 1.223 * 800,000 = 978,400.

Hence, to hedge a value of INR 800, 000, one would need to short futures worth INR 978 400.

If the NIFTY futures prices are 9025, the lot size of 25 gives rise to the contract value = 9025 × 25 = INR 225, 625.

The number of lots (N) of NIFTY Futures will be equal to

$$N = \frac{\text{HedgeValue}}{\text{ContractValue}} = \frac{978,400}{225625} = 4.33$$

In practice, fractional lot sizes are not achievable.

After employing the hedge, assume the NIFTY goes down by 500 points (or 5.5%). Let's calculate the effectiveness of the portfolio hedge and assume one can buy 4.34 lots of NIFTY futures (Table 6.4).

Table 6.1 Long and Short Futures Payoffs

Stock Price	Net Payoff	Short Futures Payoff	Net P&L
2500	2500 – 2284 = + 216	2285 – 2500 = –215	+ 216 – 215 = + 1
2290	2290 – 2284 = + 6	2285 – 2290 = –5	+ 6 – 5 = + 1
2200	2200 – 2284 = – 84	2285 – 2200 = + 85	TRUE

Table 6.2 Nature of Betas

Beta	Interpretation
$\beta < 0$	stock price responds in the opposite direction
= 0	stock is neutral and independent
$0 < \beta < 1$	stock responds in the same direction partly
$\beta > 1$	stock responds in the same direction as the markets
	the high beta stocks

Table 6.3 Value of a portfolio

Sl No	Name of the Stock	Amount	Beta (β)
1	ACC Cement	INR 30,000	1.22
2	Axis Bank	INR 125,000	1.4
3	BPCL Oil	INR 180,000	1.42
4	Cipla Phama	INR 65,000	0.59
5	DLF Real Estate	INR 100,000	1.86
6	Infosys Software	INR 75,000	0.43
7	L&T Ltd	INR 85,000	1.43
8	Maruti Suzuki Automobile	INR 140,000	0.95
Total		INR 800,000	

Table 6.4 Calculation of P&L

Decline in Value	00 points	Decline in Market	5.50%
NIFTY value	8525	% Expected Decline in Portfolio	= 5.5% × 1.233 = 6.78%
Number of lots	4.34	Expected Decline in Portfolio	54,240
P & L = 4.34 × 25 × 500 =	INR 54,240		= INR 54,240

The net result of the gain in the short position is INR 54, 125, and the decline on the long portfolio is INR 54, 240, the difference of which is too small to ignore.

If a stock ABC does not have a futures contract. For example, South Indian Bank does not have a futures contract. How to hedge a spot position in ABC stock?

Yes, one can hedge against NIFTY index futures for stocks that have no futures. For example, assume the value of South Indian Bank worth INR 500 000. If the stock has a beta of 0.75, the hedge value is 500000 × 0.75 = 375,000.

In the above example, if the spot position is too small, say INR 50 000 or INR 100 000, will it be to hedge against such positions? One will not be able to hedge against small positions, which is lower than the value of NIFTY. The remedy is to employ options contracts in place of futures contracts.

The dual benefits of hedging include price risk that can be purely hedged using futures and the benefit of leverage in the form of margin trading. Therefore, to hedge against volatility ahead of key events, it is always better to remain hedged with a hedging cost. For a portfolio of stocks, rather than one stock, the hedging is undertaken using the index futures rather than stock futures.The trader matches the maturity holding period to the expiry period of the future. Future contracts are standard instruments, and the investor's horizon may not match with the exchange. For example, the round trip cost of selling stocks and re-purchasing them at a later date could cost about 0.1% of the value of the portfolio. In contrast, the cost of buying and canceling a futures contract could be a maximum of $20.00, which is much lower.

6.3 Basis Volatility

The basis is the difference in the spot price and futures price of a futures contract. At date t, we have

$$b(t) = S(t) - F(t) \quad (6.1)$$

Under the inventory arbitrage assumptions (Smith and Stultz 1985), the futures price is expressed as

$$F(t) = S(t) + C(t, T) - k(t, T) \quad (6.2)$$

C (t, T) is the net cost of carrying, the sum of borrowed interest expense, and fixed costs net of dividends. $k\ (t,\ T)$ is the capital gain during the period $[t,\ T\ ]$. This equals basis risk as;

$$b(t) = k(t, T) - C(t, T) \quad (6.3)$$

Let $\Delta b\ (t)$ denote the change in the basis over time Δt, and let $\Delta k\ (t,\ T)$ and $\Delta C\ (t,\ T)$ denote the related changes in the capital gain and cost of carrying that gives

$$\Delta b(t) = \Delta k(t, T) - \Delta C(t, T) \quad (6.4)$$

Although the behavior of spot and futures prices displays volatility, the difference between the two in 6.4 is lower in variations. The volatility of the basis is lower than the volatility in the spot or futures price, which is the benefit of hedging.

Buying options with long tenors 12 months is easier to hold than rolling over one-month, two-month, or three-month puts. However, holding a 12-month put can end up being more costly than rolling over short-dated options over the same period. Rolling Hedges are implemented if the holding period is longer than the delivery dates of active futures contracts; the hedger can initiate a *roll-over strategy.* This involves closing out a futures contract before the delivery month and undertaking a new position in a futures contract with a longer delivery date.

6.4 Hedge Ratio

The hedge ratio, β, is the squared correlation coefficient between the spot price and futures price. The correlation coefficient (ρ) of the futures prices with spot prices is given as

$$\Delta S(t) = \alpha + \beta \Delta F(t) + \varepsilon(t) \quad (6.5)$$

Beta is the estimator for the hedge ratio and is a necessary test for hedge effectiveness.

The basis error ($\sigma_{\varepsilon}{}^2$) explains the variability.

The effectiveness of the hedge is given by ρ^2 where

$$\rho^2 = 1 - \frac{\rho_{\epsilon}^2}{\rho_s^2} \tag{6.6}$$

6.5 Options

Options allow investors to hedge their risks with a call or put option. The buyer has no obligation to pay the premium to buy or sell shares at a given price at expiry (Black and Scholes 1973). An option is the buyers' right to purchase shares at the strike price in the money. The call option holder benefits from the price rise. If the price falls, the holder loses only the option premium amount paid. The holder of a put option benefits from price fall with minimal risks in the event of price rise because the owner loses the option premium price alone.

Call options are exercised on the expiry date when the spot price (S) is higher than the exercise (strike) price (K). Options are denoted in the money, at the money, out of the money, etc.

A call option is in the money (ITM) so long as $S > K$, at the money for $S = K$, and out of the money (OTM) while $S < K$. A put option is in the money (ITM) so long as $S < K$, at the money for $S = K$, and out of the money (OTM) while $S > K$. Owner exercises the option when it is in the money (ITM), to be executed on or before the date of expiry.

The classes of options include a long position in a call option, a long position in a put option, a short position in a call option, and a short position in a put option, respectively.

Option Expiry

The long-term stock options trade in three quarterly expiries (March, June, September, and December cycles). The series denote them for the month in which the expiration date occurs. Stock options are available in 3 months cycle, namely near month, next month, and far month, respectively. The executing strike prices (K) are denominated in incremental bands and spaced at intervals of \$2.50, \$5, or \$10, respectively. All options of the same type (calls or puts) are referred to as an *option class*. For example, as shown in the option chain, if a stock has three expiration dates and five strike prices, there could exist a total of 30 different contracts.

Table 6.5 Effect on the price of ·a stock option

Observed	European call	American	European put	American put
1. Current stock price	+	+	-	-
2. Strike price	-	-	+	+
3. Time to expiration	N.A	+	N.A	+
4. Volatility	+	+	+	-

For a put option, the payoff is the amount by which the strike price exceeds the stock price. In the European call option, the payoff to the holder of a short position is a minimum of (K- S_T, 0). In a European put option, the payoff to the long position holder is maximum (K- S_T. 0). In a European put option, the payoff from a short position is minimum (S_T- *K)*. Put options lose their value as the spot price rises and are more worthy with the increase in strike price. Table 6.5 shows the effect on option prices due to chănges in spot prices.

American options become more valuable as the time to expiry is close (Broadie and Detemple 1996). A downside put option is used as a hedge. For example, for a stock currently at $100, one can buy the one option of the 6-month $80 put for $1.0. He can step out at a price below $79 (=$80 -$1). Option holder does have a mix of choices to blend calls and puts added to the underlying to create positions to minimize risks. For example, the holder waits until the option's expiry and makes a profit based on the strike price. The holder gains by selling the option for a profit in the options market.

6.6 Netting of Margins

The margin account is maintained alongside a derivative account for holding derivative contracts such as futures or options. The purpose of margin is to set aside and aggregate the risks to the broker by consolidating or contra netting positions to measure the overall net risk. The result is lower margin requirements for hedged positions. The margin account displays the final balance after all positions are adjusted against each other. The benefit accrues to the derivatives traders by allowing them to take advantage of leverage.

Bilateral netting is the process of consolidating counter liabilities between two transactions or exposures. Netting is taken into account to reduce the net derivative exposure. Netting is combining trades on derivatives and cash positions, rcfcrring to the same underlying assets by similar maturity dates or different maturity dates. The netting is intended to limit, offset, or try to eliminate the overall risk.

For example, for each option contract, the premium payable amount and the premium receivable amount are netted to calculate the net premium payable from each client.

6.7 Put–Call Parity

Put–call parity defines the relationship between the put options and call options of the same underlying stock,

strike price, and expiry date. It states that holding a short put and long call of the same class delivers the same return. The put and call option prices cannot differ because an arbitrage opportunity cannot exist.

The equation expressing put–call parity is

$$c + Ke^{-rT} = p + S_0 \tag{6.8}$$

If the laundry list includes a variety of stocks; hedging every position could be extremely difficult and costly. There are examples of trading strategies with derivatives. If the stock is currently overpriced, the investor is better off selling the option than exercising it. For example, a put option, alongside the stock in hand, can insure the holder against a price drop. Similarly, a covered call sells a call option alongside buying the stock. Further, a protective put buys the stock and simultaneously buys a put option.

6.8 Hedging Strategy

The principal hedging methods are desciribed here which follows from the types of sensitivities, such as, delta, theta, gamma, vega and rho, respectively.

6.8.1 Delta Hedging (Δ)

Delta is the ratio between the changes in spot price and the change in derivative price (Ederington 1979). For example, a stock option with a delta of 0.65 implies that if the underlying stock price increases by $1 per share, the option price will rise by $0.65 per share. Delta varies between 0 and 1. Delta can be negative for a call option and (–1) to 0 for a put option. When the delta (Δ) of a call option is 0.6, it implies against a $10 change in spot price; the option price will change by $6.00 only.

6.8.2 Theta (Θ)

Theta is a negative number. Theta is the time decay or fall of the option price. It is referred to as time decay. The option loses value as time moves closer to the maturity of the option.

6.8.3 Gamma (Γ)

Gamma is the change of an option's delta (Δ) against the unit change in stock price. Gamma is the first derivative of the delta. It gives a fair idea about the option when it is in or out of the money. Gamma is the second-order derivative of the option price against the stock price. Gamma increases when the option is closer to the money. Gamma decreases when the option is within the money. Gamma also approaches zero the deeper an option gets out of the money. Gamma is at its highest when the price is at the money. For example, if a call option has a delta of 0.4, the option will increase in value by $0.40 when the stock price increases by $1. Meanwhile, due to the $1 increase, if the new delta is 0.53, the difference in the two deltas is an approximate value of gamma (=0.13).

6.8.4 Vega (V)

Vega is the change in option's price to change in volatility of the underlying asset. Vega shows the option price changes to a unit change in implied volatility. Vega rises due to price movements in the underlying stock and falls at expiry. The right stock may turn out to be the wrong stock in the lapse of time and could be in 90 days due to change in Vega.

6.8.5 Rho (ρ)

Rho is the change in the option price to the change in borrowing rates of interest. It measures the impact of the option price due to risk-free rates of borrowing.

6.9 Efficacy Tests of Hedging

The hedge effectiveness test is a forward-looking test of the hedged item's effective benefits during the term. It assesses the changes to the hedge portfolio and compares the performance of the hedge relationship. The observation outcomes will include mean returns, the standard deviation of returns, and a promising reduction in variance.

6.9.1 Qualitative Methods of Hedge Effectiveness

This is the non-mathematical test that includes the validation of a quick checklist of terms. The critical conditions allow proving quantitatively that the hedge is effective. Under the critical situations, the stock and hedging instrument is perfectly matched to conclude that the changes will exactly offset. When the hedging plan is fully (100%) effective, it can be stated that it is expected to be effective in the future. The notional value of the derivative is equal to the notional value of the stock. The maturity of the derivative equals the maturity of the stock exposure.

6.9.2 Sample Window

The sample design and sample window are parameters of tests of the effectiveness of hedging. In laymen terms, as mentioned in Chap. 5, a portfolio gets better when the standard deviation falls due to the addition of uncorrelated stocks to a portfolio. The outcome could be tested ex-ante or ex-post depending on whether the hedging needs evaluation before or after the derivative position. The tests with or without treatment should result in statistically distinguishable changes to the volatility. The dataset is split into two parts, the training sample and the test sample, respectively. The training sample creates the hedge ratio, and the test sample assesses the performance of the hedge ratio.

The hedge ratio is tested for its stability so that the hedge remains effective.

The regression model can provide an estimate for the hedge ratio. To account for autocorrelations in prices, the *rates* of change in prices are fitted against the rate of change in futures as follows:

$$\frac{\Delta {S_T}^2}{S_T} = \alpha + \beta \frac{\Delta F_T}{F_T} + \epsilon \tag{6.7}$$

The slope β captures the change in spot relative to the change in futures. The hedge ratio is the multiplication of β *and* spot-futures price (S_T/ F_T).

6.9.3 Split Tests of Hedging

In the split tests, the post-period sample of hedge ratios distinguishes between one or more treatments or strategies. For example, the daily data on the bank Index and the NIFTY Futures indices are considered for the period from June 1, 2015, to December 31, 2016. The entire period is broken down into pre-period from June 1, 2015, to August 31, 2015, for evaluating the in-sample hedge ratio and the post-period from September 1, 2015, to December 31, 2015, for the hold out sample tests. The results for both the in-sample and post-sample performances are calculated to visualize the hedging effectiveness. The daily data is grouped for multiple horizons from 1 day, 5 days, 10 days, 15 days, and 20 days. The multiple horizons could control for the average effect of time windows. Each hedge horizon represents a different hedging strategy on the same sample.

The post-treatment observation outcomes will include mean returns, the standard deviation of returns, and the reduction in variance which is promising. The outcomes are observed against hedge ratios for comparison consisting of the mean or variance of returns. The hedging treatment is considered effective if the mean (post hedge) returns are higher. The hedging treatment is effective when the (post hedge) variance is lower than no treatment (unhedged). Hence, the percentage in variance reduction is the trade-off between risk and returns.

6.9.4 Model

In the case of split sample tests, between the treatment sample and control sample, the degree of influence of the hedging strategy could differ. This is because many other factors may control the results. The DID model distinguishes the outcomes between treated and control samples over time. The difference-in-difference (DID) model is recommended when one of the samples could experience a treatment against other samples, which are not subjected to any treatment. This is referred to as the "common trends" assumption, with more than one period of pre-treatment observations.

DID model is given as,

$$Y_i = \alpha + \beta_1 \text{Time} + \beta_2 D_i + \beta_3 + \beta_4 X_i + \varepsilon \tag{6.9}$$

where Y_i is the outcome of the treatment (e.g., returns, standard deviation of returns, and reduction in the variance), Time is the period of observation or the duration of hedge,

D_i is the choice of treatment (e.g., Strategy 1, Strategy 2, etc.),

X_is are specific situations in the strategy (e.g., Hedge ratio, combinations, etc.).

The DID model is analyzed using a regression model.

The Limitations of DID include the need for baseline data or a permanent control group. In practice, any base year or base period, which is a normal period, can be chosen as a control.

The problem of endogeneity or joint effects are eliminated. Earlier tests include the significant difference in hedge ratios between high- and low-leverage firms when leverage drives hedging intensity.

Limitations of Tests

In the endogeneity concern, propensity matching is employed, whereby each treatment group is matched with an observation from the control group by relevant attributes. The difference-in-difference tests are repeated by restricting the sample to one over another.

6.10 Volatility Strategy

As discussed in Chap. 2, past volatility can help predict future short-term volatility. Due to the nature of the distribution of volatility, the "mean exceedance" of the losses, which is the right tail, can cut down the overall losses due to the volatility strategy undertaken. The volatility strategy reduces the likelihood of extreme returns. To introduce the idea, when one investor is sensitive to volatility, he could construct a portfolio that would combine the tough challenges of the total variance, expected return and market variance, and specific risk that exists due to one or more correlations among the constituents. For example, all portfolios that give sub-optimal returns are bound to have lower risks. This leaves him to choose between (1) minimum variance portfolio, (2) target volatility portfolio, (3) equal volatility-weighted portfolios, and (4) others (e.g., equal capital weighted, equal Sharpe ratio, etc.) which are described hereafter.

6.10.1 Volatility Derivatives

Volatility-based derivatives are a class of derivatives where the payoffs depend on measured degrees of the price deviation in the underlying asset. These derivatives include variance swaps, IX futures, VIX options, etc. VIX (Volatility Index) is an option determined expectation of market volatility over a period. Volatility Index is the distance by which the index is expected to change derived from the options order data. India VIX reports NIFTY Options contracts to indicate expected NIFTY volatility over the coming month of 30 days. Volatility Index is a measure of the amount by which an underlying index is expected to fluctuate in the near term (calculated as annualized volatility) in percentage. India VIX is a volatility index based on the NIFTY Index Option prices over the next 30 calendar days.

6.10.2 Risk Parity

Risk parity is an allocation mix of a portfolio based on "riskiness". It follows the modern portfolio theory (MPT) method to decide on allocations. The end goal of the risk parity technique is to generate optimal portfolio returns at a target level of risk. For example, the weights are calculated by target risk level instead of a pre-determined proportion of asset diversification (equity and fixed income (60:40). It allows managers to choose specific risk levels and achieve diversification. The risk goals are realized using varying levels of leverage, assuming approximately similar among stocks. The individual capital share of each stock is different from each other, although they result in almost identical risk levels.

6.10.3 Equal Volatility Weighting

It is a risk-weighted approach that is unequal to market sensitivity (capitalization). Instead of allocating capital to assets based on market sensitivity (capitalization), equal volatility weights the securities based on total individual risks. It assigns lower weights to higher risks and gives more weight against the lower total volatility of the asset. The result of the approach makes the contribution of every asset to overall portfolio volatility equal. It does not guarantee equal capital across assets.

The equal volatility formula is given as:

$$W_i = \frac{1}{\sigma_i} / \{ \frac{1}{\sigma_1} + \frac{1}{\sigma_2} + \frac{1}{\sigma_3} \ldots . + \frac{1}{\sigma_j} + \ldots \} \quad (6.10)$$

The limitation is when the volatility of a security is unusually low, larger weights are assigned to it. When the volatility falls back to a previous normal level, the weights are too large to be reviewed. It does not take into consideration correlations. The distinction between equal volatility and risk parity lies in considering the bivariate correlation of two induced assets into the basket.

6.10.4 Volatility Targeting

Volatility targeting is used to determine a given degree of leverage in a portfolio to maintain given total volatility. The manager alters the proportion of leverage to match the target volatility of the portfolio. When the volatility falls, the leverage is enhanced. The more volatile the portfolio and the higher the amount of leverage used, the riskier the portfolio. The equal-weight portfolio performs better than the volatility-weighted portfolio on similar counts.

If one considers the volatility of a base (unlevered) portfolio is 3% per month (10.39% annual). To reach an annual volatility of 20%, the leverage has to have arrived at 1.92, which means that, for every dollar invested, an additional $0.92 is borrowed to meet the target volatility of an annual 20%.

6.10.5 Zero Variance Portfolios

Portfolio variance is the weighted variances of each stock in a portfolio. The portfolio risk falls when there are non-correlated stocks. The increased idiosyncratic (specific) volatility of individual stocks implies that it is more difficult to diversify away idiosyncratic risk with only a limited number of stocks in a portfolio. Firstly, it seeks to reduce losses by reducing the total volatility. Secondly, the creation of portfolios aimed to capture the "low volatility effect". In minimum variance, the weights with the lowest expected risk are used. This may include the possibility of overly concentrated exposure to the sector, country, or individual stocks. Two assets with a low correlation can be the candidates of a minimum variance portfolio.

6.10.6 Equal Weighting

An equal-weighted index fund invests in equal capital proportions. An equally weighted index, for example, puts the same amount of money into Apple as it does into American Express. It puts the same amount of money into companies as in the index, regardless of their size. The equal-weight and volatility-weighted portfolios are not the same. The volatility shares are aimed at matching the portfolio volatility limits. The volatility weighting is reviewed quarterly. The manager uses the macroscopic and passive approach to construct an index for each country with stocks. Low-volatility stocks have higher weights than others. The new stocks are selected based on the correlations with a global portfolio. The results are a diversified portfolio and a universe of many stocks.

6.10.7 Global Portfolio

The global portfolio involves many asset classes, including stocks, bonds, commodities, and alternative assets with largely uncorrelated and diversified returns. It is based on risk parity that enables investors to benefit from uncorrelated factors of market beta. It uses a quantitative method to arrive at allocations than a simplified allocation. Reallocation could mean revising the weights against an alternate stock that belongs to the same sector as the existing member. It diversifies investments across both the industry sector as well as market capitalization and geographic region. It may not be practically possible for all classes of investors. One should hold all asset classes, which comprise gold, oil, bonds, sovereign papers, and bits or slices of small, medium, and large stocks, all of them inside an ideal basket. Inside a portfolio, within each sector, the correlation between stocks may be too high. Risk parity is a discipline preparing a basket of sectors and geography and large caps and small caps. Exposure to global currencies in reasonable proportion would mean the construction of a global portfolio. Risk parity may produce a better risk/reward trade-off than capital-allocated baskets. The limitations of risk parity are the illiquid stocks (or constituents) across underdeveloped economies.

For example, FPIs in India invest up to INR 1.5Lakh crores in VRR (Voluntary Retention Route) Debts as per RBI limits for a minimum period of three years. The FPIs in India routinely invest in direct Equity, Primary market equity, Secondary Debt, Hybrids, Primary market equity, etc. Further, they also invest in derivatives or ETFs in the form of Index Options, Stock Futures, Stock Options, Interest Rate Futures, etc.

6.10.8 Volatility Targeting

Volatility targeting is a method to reallocate assets within a basket based on their volatility. It is almost similar to switching from low beta stocks to high beta stocks. The target volatility level can drill down to high-performing stocks that could have had the lowest drawdowns. For example, when the index has realized volatility of annual 15.3%, the volatility target of two-thirds of the index at an annual 11% is fixed. Long-horizon investors who are less concerned with downside protection may favor a higher volatility target to better meet return objectives. Conversely, risk-averse investors seeking downside protection prefer lower target volatility. When the index undergoes a drawdown of 50%, which is brought down to one-third (33%) using the targeted volatility approach.

6.11 Summary

This chapter elaborated on hedging instruments. It illustrated examples and explained why firms choose to hedge certain types of risk. It briefly described risk-based netting, risk parity, volatility weighting, and volatility targeting between the equal volatility and volatility-weighted portfolios. The equal-weight portfolio performed over the volatility-weighted portfolio. Most importantly, sampling methods and the ex-ante and post facto tests of the efficacy of hedging were also discussed. The efficacy tests' credibility depends on how well to address and verify the accuracy related to the effectiveness and useful guidance on goals. We explained minimum variance, hedge ratio, and sampling and split tests and their limitations.

6.12 Questions

6.11.1 What is target volatility? Is it the same as Realized volatility?

6.11.2 How do you conduct an efficacy test for small trading history?

6.11.3 How will compare between the results of the test of two different hedging solutions?

6.11.4 What is split sample testing?

6.11.5 Mention the six different positions an investor can follow when he expects the price of a stock to rise.

6.11.6 Explain the variation of an investor's net payoffs when the call option has an exercise price equal to the current stock price, with the expiry price for two positions with the following:

(a) one short call option and one share;

(b) one short call option and two shares;
(c) two short call options and one share;
(d) four short call options and one share.

6.11.7 Give an example and explain equal volatility weighting and its advantages.
6.11.8 A protective put is created when a long position in a stock is married to a long position in a put on that stock. What position in a call option is almost equal to a protective put?

Solutions

6.11.1 Yes. Because the target volatility is calculated only ex-post.
6.11.2 There is no shortcut to sample preparation. Thin trading history must be made into sufficient sample size by adding similar hedge profiles of related underlying assets in the same situation.
6.11.3 The outcomes are observed against hedge ratios for comparison including the mean or variance of returns. The hedging treatment is considered effective if the mean (post hedge) returns are higher. The hedging treatment is considered effective when the (post hedge) variance is lower compared to no treatment (unhedged). Hence, the percentage in variance reduction is the trade-off between risk and returns.
6.11.4 The split sample test includes cohorts or stratified random splits of one or windows. The splitting of the sample leads to the same indicator value for each subsample, for comparison, against the indicator's value for the full sample or close to the full sample.
6.11.5 There are six moves, namely (i) long position in stock and a short call, (ii) long position in a call and short position in a stock, (iii) long put position and long position in a stock, and (iv) short put position and short position in the stock.
6.11.6 Let SBI be bought at INR 320, and in the next 3 days, the stock went up to INR 375. One option is to book profits at INR 375 and exit SBI. But, due to positive news, the stock may reach INR 450 in the next three months. One hedges the profits by selling SBI Futures at INR 375 and realizes INR 55. If the stock of SBI crosses another resistance, roll over the futures to the series of the next month. This is the equivalent of a trailing stop-loss in a profitable situation.
6.11.7 For example, let the annual and realized volatility be 4%, and 6% respectively. The asset weights are multiplied by 4/6, and the balance is allocated to a risk-free asset. This strategy aims to sell stocks when their expected return falls due to volatility. It intends to buy stocks when their

expected return is rising due to volatility. In this volatility weighting-based allocation method, the risk-return ratio does not improve compared to an equal-weight portfolio.

6.11.8 A protective put is the sum of a long put position along with a long position in underlying. Hence, it equals a certain amount of cash and a long position in the call option. This follows from the put–call parity relationship.

References

Black, F., and M. Scholes., 1973. "The Pricing of Options and Corporate Liabilities," *Journal of Political Economy,* 81 (May/June 1973): 637–59.

Broadie, M., and J. Detemple. 1996. 1996 "American Option Valuation: New Bounds, Approximations, and a Comparison of Existing Methods,." *Review of Financial Studies* 9 (4): 1211–1250.

Ederington, L.1979. "The Hedging Performance of the New Futures Market", *Journal of Finance,* Vol. 34, March 1979, 157 − 170.

Merton, R. C. 1974. "On the Pricing of Corporate Debt: The Risk Structure of Interest Rates," *Journal of Finance,* 29, 2 (1974): 449–70.

Merton, R. C. 1973. "Theory of Rational Option Pricing," *Bell Journal of Economics and Management Science,* 4 (Spring 1973): 141–83.

Smith, C., and R. Stulz. 1985. The Determinants of Firms Hedging Policies. *Journal of Financial and Quantitative Analysis* 20 (1985): 391–405.

How Far the Risk

7

Learning objectives:

- Determine how far to bear the risk
- Determine how much to lose per exposure
- Give a Microscopic view of the loss process
- Derive the end impact of eventual losses
- Provide guidance on carrying forward losses, obtaining tax claims
- Familiarize the students with the alternatives to handle loss carrybacks
- Introduce benchmarks of global portfolios

7.1 Microscope

Risk is not limitless. The art of constructing a portfolio can pull or push an investor in or out of the market. The distinction between microscopic and macroscopic landscapes is the ability to beat the consensus. Telescopic view refers to focus on an overall bigger picture and a macro view of the market. It is pursued with a top-down approach. The telescopic view helps zero down the stock. While telescopic view can mean following the herd, the microscopic view is a pure independent attempt to beat the consensus.

A top-down approach is also known as an EIC (Economy, Industry, and Company)-based approach. The process flow of top-down investing follows the stronger macroeconomic situation in terms of higher expected growth, low inflation, and low-interest rates, among others. Is the industry situation conducive to outperform? What is the demand situation, unmet market needs, etc.? Does the company have intrinsic strengths in terms of profits and solvency? What are the operating margins, efficiency ratios, and the valuations of the stock? The top-down approach works when the basic approach to investing is focused on large-cap

D. Bag, *Valuation and Volatility*,
https://doi.org/10.1007/978-981-16-1135-3_7

stocks. In any market, the large-cap stocks tend to be more vulnerable to macro factors than the smaller companies. For example, when the interest rates move up, the large rate-sensitive stocks get impacted more than the insensitive ones.

The microscopic evaluation of stocks is related to a trend of the market index from years of weekly or monthly historical data (Shabacker 1930). Later, the daily return for two years or a lower time frame, such as one month or 20 days for six months, is observed. The choice of the sector is also important in top-down or bottom-up approach. In industries like banking, commodities, and autos, the macro factors play a bigger role for the top-down approach to work better. On the other hand, for sectors like pharma, auto ancillaries, software, etc., the micro factors play a much bigger role. In these cases, it is possible to differentiate at a company level, irrespective of the macro environment. Global institutional investors prefer the top-down approach as it provides a context to invest. For example, India-specific funds are driven by macro and industry factors since they benchmark to the MSCI EM or similar index. There is a lot more value in bottom-up investing for an individual investor, a portfolio management scheme (PMS), or a domestic mutual fund. Therefore, both approaches are complementary than competitive. The bottom-up approach is best when the market conditions and the macro conditions are normal. An investor uses the bottom-up approach to identify the right stocks to invest in and apply the top-down approach to time his entry and exit. The microscopic approach believes that good companies are good investments irrespective of whether the overall economy is good, bad, or ugly.

7.2 Exit a Trade

Many traders enter a trade without an exit strategy and run into losses. What are the resorts available to help minimize losses? The decision to exit a position is psychological or specific to the trader. There exist examples where a well-defined exit strategy can improve a random entry into a trade. Many traders may not specialize in exit techniques (LeBaron Blake 1999). For instance, stop-loss is the price at which a trader will exit selling the stock and a loss on the trade. Stop-loss orders are orders to sell a stock the moment a price point is achieved automatically. At the pre-determined price point, the stop-loss is converted into a market order. The stop-loss expires after one trading day. In situations of a market correction, when a correction appears with a decline of 10% or more in the price of a stock from its recent peak traders.

For example, assume one has bought a stock at an entry price of 50 per share. Now the stock has lost 10% of its value to currently trading at 45. If he decides to exit the trade to prevent making further losses, the stop-loss limit would be set at 45. He loses 5 under the current market value of the stock ($50 \times 10\% = 5$).

They could follow a narrow stop-loss of 1% in a single trade. For example, for a position of $10,000, the stop-loss limit is less than $100. Moving averages of the prices are gauged to set the limits.

For example, a limit of loss for a trading account could be is 2% of the owner's capital. Traders understand and evaluate their risk tolerances to determine stop-loss placements. Many traders prefer losses to be limited to 0.25–0.5% of their individual exposures. Institutional traders are more risk-averse than self-proprietary traders. When the stop-loss order is executed at 10% below the purchased price of the stock, for example, if a stock is purchased at $30 and the wider stop-loss is intended at 20%, the loss limit is to be priced at $24 only.

There are two kinds of stop-loss, the simple and the trailing stop-loss. Trailing stop-loss follows rising prices, where the stop-loss is fixed at a percentage of the current market price rather than at an absolute amount. The trailing stop limit is revised upwardly. The moving average of prices guides the stop-loss that could be placed below a longer-term moving average. A longer-term moving average is assigned for more volatile stocks. Alternatively, stop-losses could be 1.5 times or a fixed multiple of the current high to low range. A price target is a price to which a stock is expected to rise. If the stock price remains stable, then stop-loss needs to be compact. The stop-loss and target prices shown in public TV channels and media networks altogether vary for individuals or will not match the safety needs of institutional players. Similarly, the SMS alerts received live from local brokers that provide real-time target price and stop-loss price are not accurate or reliable.

As a matter of discipline, making routine revisions to the stop price is mandated (Edwards and Magee 1997). Stop-loss is an art since it also depends upon the leverage limit availed by the borrower. The stop-loss limits move closer to the magnified loss limits driven by margin trading. For example, institutional players may fix the stop-loss limit further below the target price. Institutional players may expect higher target prices. Wise Traders make periodic revisions to limits on how much they are willing to win or lose. Wise traders periodically assess their performance, review their positions, and never repent.

7.3 Ordinary Loss

The loss in business is termed ordinary loss by a taxpayer when its expenses are more than its revenue earned during the course of its operations. Ordinary losses are incurred by the entity, which is not termed capital losses (Bunea-Bontas 2009). An ordinary loss is affected and deducted in full to set off the reported income. The deductions of losses reduce taxes owed by a taxpayer. Ordinary losses may stem from damages, disaster loss, fire, transit loss, theft, etc. A business can claim a deduction for interest expenses, the interest expended on funds borrowed to purchase investments of taxable nature. This includes margin loans for buying stock in the trading account. The interest on the margin loan is deducted. During tax deduction at source (TDS), the broker deducts service tax from the net gains after from deducting the interest costs from margin funds. Ordinary losses are not exactly capital losses. The realization of a capital loss happens when a capital asset is sold, such as a stock market investment or property. A capital loss could be fixed to a limit of setting a capital gain of only $3,000 (three thousand) of ordinary income.

7.4 Tax Benefits of Loss

How far can an investor go? How much taxes can he save out of overall losses made? Stop-loss limits are based on how much tax savings would accrue when the trigger is missed. Assuming the assessee is in the 30% income tax bracket, he pays back 10% of the gain as CGT and files for paying cumulative taxes @30% (excluding surcharge) to the government. He retains less than 70% (could be lower) of his total gains. Conversely, when the taxpayer makes a capital loss, he saves on 10% of CGT, and further, he can adjust the losses in total tax to be credited to the government. In the same bracket of 30% (excluding surcharge), the net loss is less than 70% (could be lower) of his net gain. The net gain on tax amount serves as the boundary for the player, which comes to his rescue on the top of the stop-loss limit.

7.4.1 Long-Term Capital Gains

Long-term capital gains are derived from investments that are held for more than one year. A short-term capital gain from security held for a period below one year or less is treated as ordinary income. The gains in the short term are treated with one to seven (1–7) tax brackets. The taxable income limit exceeding $434,550 (single or married and filing jointly) is subject to that highest rate (@) of 37%. The long-term gains are subject to rates (@) of 0%, 15%, 20%, or 37%, respectively. Short-term gains on listed securities are taxable at 15% in India. The single filter short-term CGT stands at 32% in the USA. Short-term gains are taxed as regular income according to tax brackets up to 37%. Long-term gains are subject to rates of 0%, 15%, and 20%, based on income limits. In India, it is 10% for the long term and 15% for the short term. In a few situations, TDS of 10% may be applicable at the source, such as income distributed by Fund Houses applied at the source. The receiver would pay taxes calculated after receiving the net income of TDS, deducted at the source.

Assume a salaried professional with a salary income of 200,000 and indulges in periodic trading after holding stocks for less than one year in the last financial year. He had made a short-term capital gain (STCG) of 4000 and had also made a loss of 1000. Here, the short-term capital loss cannot be set off against his income from salary or similar sources. However, the short-term capital loss can be set off against capital gains only to the limit of loss which is 1000. The balance of 3000 (=4000–1000) will be carried forward. Therefore, his taxable income remains at 200,000.

Table 7.1 The arrival of the long-term capital gains

Sale Value		95,150
Less:	Indexed cost of acquisition of an asset	78,960
Less:	Indexed cost of asset improvement or expenses incurred in the transfer of the assets (including commissions, brokerage, etc.)	0
Less:	Expenses incurred concerning transfers or sale (includes costs incurred post asset purchase for the improvement of the same)	0
Long-term capital gain		16,190
Net Gain		3238

A recognized loss occurs when an identified asset is sold for less than its purchase price. It is reported for income tax purposes and is carried over into future periods. The recognized loss is deducted from capital gains. It enables individuals or entities to reduce their tax outgo bills. A loss is realized immediately after an investor completes a transaction. Although a sale creates both discovered and recognized losses simultaneously, only a recognized loss may be deducted from capital gains. The exceptions of the Wash Sale Rule, where, once an investor incurs a loss, the same stock or an identical stock or fund cannot be bought back for 30 days following the sale.

Table 7.1 shows the arrival of capital gains tax on an investment of $ 28,200 made in listed stocks 5 years ago, which resulted in a sale value of $ 95 150. The indexed cost of acquisition (inflation) is given as $(=\frac{280}{100})$ 2.8 times the purchase value $= 2.8 * 28,200 = 78,960$. Assuming there are no improvement costs, and the taxable gain equals $16,190; Applying the tax rate @20%, the net gain is only $3238.

7.4.2 Carryforward Losses

A carryforward (or carryover) of losses is a provision that allows a tax loss to be carried over to future years. The tax loss carryforward can be claimed by an individual or a business to reduce future tax payments. Capital losses higher than capital gains realized in a given year may be used to set against ordinary income (taxable) for a limit of $3,000 in one year. For tax purposes, realized capital losses are used to reduce the tax outgo only if the asset sold was owned for investment purposes. The capital losses are divided into two categories. Short-term losses occur when the stock is disposed of in less than a year. Long-term losses occur when the stock has been held for a year or more. The capital loss is the sale value minus the adjusted purchase cost for the number of shares sold. The adjusted amount includes all-inclusive admissible transaction fees, brokerage, commission, charges, margin costs, etc.

Table 7.2 Carrybackward of net operating losses to previous years in 2015

Operating loss amount	Carried amount	Year
$10,000	$0	2015
$4,000	$6,000	2013
$0	$4,000	2014

7.4.3 Loss Carryback

Loss carrybacks are synonymous with loss carryforwards. Loss carryback is attempted when the business realizes a net loss and intends to apply it to reported prior gains. This reduces the liabilities for previous years. The loss carryback can result in tax refund accruing to previous periods. In the USA, losses are carried back only two years prior to the year in which the net operating loss occurred. A business may choose to carry the loss forward if it expects increased tax liability in the future. In India, it can be carried forward for a period of 8 years and adjusted against any short-term or long-term capital gains made during these 8 years.

Table 7.2 demonstrates the feature of loss carryback. One goes back to the basis year, which is 2 years prior to the loss year, which could be the current year 2015. For example, if one has an operating loss in 2015, he goes back to 2013. He effectively recalculates the basis year 2013 liability for tax after deducting the volume of operating losses. Similarly, suppose a part of the operating losses remains unset even in two years prior. In that case, the remaining NOL is subtracted from income in the last basis year, one year prior to the operating loss year.

7.5 Market Triggers

Market triggers are signals for making decisions; specific examples of many of which are discussed in use cases in Chap. 8 of this book. There are relative assessments of market positions devised from the relative stands of optimistic counts against pessimistic counts. Although practitioners are tempted to use a variety of modern triggers, there are three or more major conventions such as (1) change in the volume of open calls or open puts, (2) Advance/ Decline Ratio, and (3) Put–Call ratio, respectively.

7.6 Support and Resistance

The support and resistance are the levels of price of a stock. The exchange never displays such price levels. It exists apparently and is psychological in nature. One trader may interpret it differently from one another (Sweeney 1986). As mentioned in Chap. 1, the goals of one trader are different from another. Support is a price level from which the stock does not fall below. The support level is the floor price of the stock. The resistance level is the roof ceiling price level of the stock. The level acts as a ceiling and prevents the stock from rising any further.

In real-time basis, as news flows, the open interest level changes. For example, the volume of open interest is used as an indicator for the same. For any given stock, the daily option chain displayed by the exchange includes activity, open interest, and price changes for each underlying option. The indicators for support and resistance are the highest or maximum displayed values of open interest of call and put option. Open Interest (OI) shows the total number of options contracts that are currently outstanding (open) in the market. It indicates the level at which traders have built positions expecting the market to go up or down. The large interest in open calls acts as options-related resistance. Larger open put interest acts as options-related support. A long position is buying a stock that may rise in two months. Open interest shows the total number of outstanding options that are yet to expire or settle. The higher the open interest, the higher the level of interest in a position. Short covering is buying of underlying stocks to cover up an existing short position.

Short covering is necessary to avoid loss on a short position when prices start moving upward. Short cover signals a shift from bearish phase to bull phase. The sellers of put options may sell the underlying stock short.

For example, the price rise and rise in open interest would mean new money enters into the market or new buying, which is bullish. If the volume of open interest has fallen off and spot prices are declining, day traders are liquidating their positions, and a bearish phase is gathered. A recently observed higher open interest at a market top and subsequent price fall off are considered bearish. When the price rises and opens interest declines, the short-sellers cover their positions and give a buy signal. During a price decline, if the open interest remains flat, it gives a sell signal. In long positions, the larger the volume of open interest, the higher the level of bullishness. Short covering refers to buying of shares in order to close an existing short position or a position that has been sold. Typically, short covering is done to avoid loss on a short position when prices start moving upward.

When the options expire or the put buyers unwind their positions, the short interest will be repurchased, which can add to the buying pressure as the underlying shares reach closer to prices.

7.7 Advance-Decline Ratio

Support and resistance levels depend on the advance-decline ratio (A/D) that compares the volume of shares that ended higher in contrast to the volume of shares that ended lower than the previous closing prices for the day. The advance-decline ratio is the ratio of the number of advancing shares to the number of declining shares. The A/D ratio is a tool that helps traders observe trends and likely reversal. The advance-decline ratio is calculated for one day, one week, or one month, etc.;

the advance-decline ratio can signal that the market changes directions. A low ratio value indicates an excess supply, or a higher value of the ratio indicates an excess demand. A steadily increasing ratio might signal a bullish trend, and the opposite would indicate a bearish trend.

7.8 Put–Call Ratios

The ratio represents a proportion between the total put options and the total call options purchased on any given day. A put–call ratio above 1 indicates a sale signal, and a put–call ratio below 1 is a buy signal. Few contrarian traders may believe the opposite to be true and buy when the ratio is above 1 or sell when the ratio is below 1.0. When the open interest is higher, it is a signal to decide when to buy and exit the market. Short buildup means traders expect the prices to go down and create short positions. Short covering is known as buying cash or spot to cover up the sell positions, which is spot buying in order to close out an existing short position. The short-selling transaction will be covered when the spot purchase is equal to the number of sold stocks.

7.9 Motivation and Response to Trading

There exists bias or prejudices on the part of traders. The psychology of trading refers to the mental state and emotions that result in success or failure in trading. The characteristics of an individual's behaviors influence his motivation to trade during trading actions. It is as important as other characteristics of skills, experience, and knowledge, respectively. Discipline, self-control, and perseverance are needed (Neter and Waksberg 1964). Fear and greed are the emotions associated with trading or hope and regret. Greed drives to accept risk. Fear drives to avoid risk.

Greed is the desire for more wealth that hides rational judgment. Greed often leads toward making high-risk trades. Greed may motivate traders to hang on to profitable trades longer than real to seek profits. Fear motivates traders to escape away from positions prematurely. Fear morphs into panic and sells off from panic selling. It is an overreactive emotion that causes traders to act irrationally and exit the market. To conclude, investors bitten by the bear market would deploy a compounding principle against the cause. If the bear market drops 20% in value, a 10% rise will return the portfolio to 88% of its original value. Therefore, it is just a question of time.

7.10 Summary

There are no concrete barricades on risk tolerances and boundaries. The previous Chap. 5 had discussed tools of assessment and fixing of limits in the dealing room using quantitative approaches. This chapter provided an overview of risk parity, Microscopic and Telescopic views, etc. It discussed ordinary losses, tax benefits, and carrying forward of losses. It provided guidance on simple benchmarks of global portfolios. It addressed the psychology of trading; the general nature of tolerances is starkly different from the classes of investors discussed in Chap. 1. The boundaries and limits for an investor would have depended upon his risk-return goals, horizon, liquidity, group policy, and regulatory norms, respectively.

7.11 Questions

7.11.1 Calculate the stop-loss and trailing stop-loss from the following data if the market price rises to $10.97.

7.11.2 Calculate the Tax-Loss Harvest from the assessee information given below who falls in the tax bracket 24% for the assessment year.
An investment of $10,000 in stocks with a leverage of 10% at the current market value of $9,000 results in a realized loss of $1,000. Later due to price rise, the initial purchase of $10,000 moves to $10,800 at the end of the year, yielding a 10% pre-tax return that also includes a 2% dividend. https://www.investopedia.com/terms/d/dividendyield.asp yield.

7.11.3 Calculate Capital Gains/Losses from the Data Below.
1,000 shares of ABC stock are sold for a capital loss totaling $10,000 after owning the stock for three years. If the taxpayer has $3,000 in long-term gains, that reduces the long-term capital loss to $7,000. Further, a loss of $2,000 occurred in short-term capital gains, leaving the remaining capital loss to $5,000. Find the carryforward amount of loss.

7.11.4 Calculate the Net Benefit of Short Covering from the Data Below.
The maximum open interest in put options stands at index level 10,800 strike prices, while the highest open interest among call options stands at index level 11,000 strike prices. The average share price at the 10,000 call option is INR 50. The average share price at the 11,000 call option was INR 83.

7.11.5 Calculate Stop-Loss for ABC Stock from the Below Data.
Let someone hold a long position on 10 shares of Tesla Inc. bought @ $315 per share. The shares are now trading for $340 each. What is the sale price?

7.11.6 Calculate the hedge ratio. Mr. Bonny owns 20,000 shares of a company. Mr. Bonny wishes to hedge the risk fully. Assuming a hedge ratio of 0.37 and a premium of 2.50 for a nearer put option, how many put options he needs to buy?

7.11.7 Calculate the stop-loss limit from the data below. How much is the stop-loss limit when the risk/reward ratio is (i) 1:2 and (ii) 1:5?

7.11.8 Answer the following if one buys a November expiration call option with an exercise price of $21.

a. If the spot price in November is $21.75. What is the profit in this position?
b. If the November call has an exercise price of $22.
c. If the November put has an exercise price of $22.

7.11.9 Which of the Following is the Riskiest in the Index Option Market if the Stock Market is Expected to Increase in the Near Future? The riskiest among all choices is attemtpting to sell a call options against a stock you do not posess. It denotes selling uncovered or naked calls.

a. Write a call option.
b. Write a put option.
c. Buy a call option.
d. Buy a put option.

7.11.10 A global equity manager is assigned to select stocks from around the world. The manager evaluates the returns to the return on the MSCI index, shown in the Table 7.3

(i) Calculate the total value added of the entire manager's decisions this period.
(ii) Calculate the value added (or subtracted) by her country allocation decisions.
(iii) Calculate the value added from her stock selection ability within countries.

7.11.11 One is interested in ABC stock options. He found that the 6-month $50 call sold for $4 and the 6-month $50 put sold for $3. Given the risk-free interest rate of 6% and the spot price at $48, how can he obtain an arbitrage advantage of the situation?

Table 7.3 Country wise manager's portfolio return

Country	Weight In MSCI index	Investor's weight	Investor's return in country (%)	Return of index (%)
U.K	0.15	0.3	20	12
Japan	0.3	0.1	15	15
U.S	0.45	0.4	10	14
Germany	0.1	0.2	5	12

Solutions

7.11.1 Price purchased = $10, Last price = $10.05, Trailing price = $0. 20.

When the market price rises to $10.97, the trailing stop will increase to (=10.97–0.20) = $10.77. If the last price drops to $10.90, the stop-loss remains unchanged at $10.77.

7.11.2 The unrealized investment gain is 8%. The dividend gain is 1.4%. The after-tax return could be 9.4% as a result of the unrealized gain. When it is sold off at a loss of $1,000, at the topmost tax slab of 24%, an amount of $760 (from a loss of $1000) is calculated in income tax savings, which gives rise to a 7.6% return on the initial purchase $10,000. The net after-tax return would be 16.6% (= 9% + 7.6%).

7.11.3 The capital losses will offset the capital gains during a taxable year up to $3,000 per year. The carryforward rules permit the taxpayer to set the loss of $5,000 toward future capital gains. For example, if the taxpayer has made a capital gain of $2,000, the entire gains can be set by $2,000.

7.11.4 The resistance is at index level 11,000, and support is at index level 10,800. The call option seller will begin to encounter loss once the index breaches 11,050 (11,000 + 50). The sellers will cover their short positions at 11,100. The put option sellers at index level 10,800 make the sellerslose 0,717 (=10,800–83).

7.11.5 The shares are now trading for $340 each. What is the sale price?
The shares are sold when the price falls at a stop-loss limit of $325.50.

7.11.6 The number of put options, n = 0.37 x (20,000/2.5) = 2960.

7.11.7 The trader purchases 100 shares at $20 and puts the stop-loss at $15 to expect losses not to exceed $500.

(i) When it is 1:2, then the loss to gain ratio is 5/10. When the price reaches $30 in future months, the trader intends to lose $5 per share to realize an expected $10 per share return.

(ii) When it is 1:5, then the loss to gain ratio is 2/10 (=$20-$18)/10. He would set the stop-loss order at $18.

7.11.8 *a.* If the spot price in November is $21.75. What is the profit in this position? Answer 0.75.
b. If the November call has an exercise price of $22. Answer. 1.0
c. If the November put has an exercise price of $23. Answer.2.0

Table 7.4 Country wise manager's portfolio performance (%)

Country	Weight In MSCI index	Investor's weight	Investor's return in country	Return of index	Weight × Return	Benchmark return	Difference	Weighted difference	Difference	Weighted difference
	A	B	C	D	BXC	AXD	B-A	D X (B-A)	C-D	B X (C-D)
U.K	0.15	0.3	20	12	6.0	1.8	0.15	1.80	8.0	2.4
Japan	0.3	0.1	15	15	1.5	4.5	-0.20	−3.00	0.0	0
U.S	0.45	0.4	10	14	4.0	6.3	−0.05	−0.70	−4.0	−1.6
Germany	0.1	0.2	5	12	1.0	1.2	0.10	1.20	−7.0	−1.4
SUM					12.5	13.8		−0.7		−0.6
(i) total value added								= 13.8–12.5 = 1.3		
(ii) value subtracted from country allocation								−0.7		
(iii) value subtracted from stock selection								−0.6		

7.11.9 Which of the Following is the *Riskiest* in the Index Option Market if the Stock Market is Expected to Increase in the Near Future?

a. refer to chapters.
b. refer to chapters.
c. refer to chapters.
d. refer to chapters.

7.11.10 See Table 7.4.

7.11.11 $c + Ke^{-rT} = 4 + 50e^{-0.06x1/2} = 52.52.$

$p + S_0 = 3 + 48 = 51.00.$

Using standard notations, the parameters values are inserted and are calculated above. The advantage of arbitrage implies a payoff of = 52.52–51.00 = 1.52.

References

Bunea-Bontas, Cristina Aurora. 2009. Basic Principles of Hedge Accounting. Economy Transdisciplinarity Cognition, 2009.

Edwards, Robert, and John Magee. 1997. *Technical Analysis of Stock Trends*, 5th ed. Boston: John Magee.

LeBaron, Blake. 1999. Technical Trading Rule Profitability of Foreign Exchange Intervention. *Journal of International Economics* 49 (October): 125–143.

Neter, John, and Joseph Waksberg ,1964. "A Study of Response Errors in Expenditures Data from Household Interviews." Journal of the American Statistical Association, vol. 59, no. 305, 1964, pp. 18–55.

Shabacker, R.W. 1930. *Stock Market Theory and Practice*. New York: B. C. Forbes Publishing Company.

Sweeney, R.J. 1986. Beating the Foreign Exchange Market. *Journal of Finance* 41 (March): 163–182.

Use Cases and Business Application 8

Learning objectives:

- Appraise students with deep dive into trade evaluation practice.
- Familiarize students with field and tools.
- Familiarize with field and business use of trading data.
- Conduct meaningful analysis and derive strategies.
- Analyze trade data to recommend impactful decisions.

8.1 Business Case

Business use cases cover examples of "HOW" in real-life situations. It could describe actions performed by managers assigned to the role and the outcome of which would be supervised and consumed by other managers in the company. This chapter provides use cases with trade data. There exists a multitude of tasks to be performed by junior and senior professionals inside the front end, mid-office, and back end of the treasury division of a business. The best practices adopted by the peers in the team are constantly reviewed for newer superior outcomes which may arise in the field. This chapter presents conventional tools and indicators applications, demonstrates valuation examples, and shows prior use cases in live situations with sample data.

8.2 Pre-requisites

Modern-day traders are called tech traders. The pre-requisites for trading include a computer or laptop. Technology evolves and changes; a computer with enough memory and CPU speed does not crash or stall. Most trading and charting software require memory and processors that are fast and up to date. One may jump between

D. Bag, *Valuation and Volatility*,
https://doi.org/10.1007/978-981-16-1135-3_8

the company's websites, broker websites, and the tools screens now and then. A wider monitor is preferred but is not a necessity. Specifically, liquidity in the account is needed for beginners initiating trades.

Real-Time and live market data are accessible. The tick-by-tick data is downloaded from broker exchange or broker websites. Except for the confidential attributes of the sources and origin of trades, volume and prices are available. The exchanges report daily Trading statistics, and historical trade datasets are available on a few websites. Many exchange the activity data on future and options as an added segment, including daily option chains and pre-open auction data. On a need basis, historical granular data may be purchased at official costs from exchanges. The testing method does involve the combinations of the sample window period, interval of price data, frequency of sampling, outcomes for comparison, choice of a champion-challenger (test and control) plans, etc. Choosing a longer sample window and smaller interval has the obvious benefits of accuracy. The sampling frequency may range from 30 to 60 min since it is practical to arrive at high-frequency data. The trading desks must have a strategy to buy or sell actions in real time; otherwise, small interval high-frequency sampling is not meaningful.

Table 8.1 describes a sample Trading Journal, which is prepared in a spreadsheet for the sake of trade review from time to time. It is the prepared and updated end of the day on a daily basis by the operator.

8.3 Tools

In general, tools are needed for analysis, spotting trends, and buy/sell decisions that require an understanding of simple measures of market movements. Traders with keen attention to changes in volatility tend to demonstrate discipline, confidence, and a balanced trading mindset that contributes to success. For example, there are indicators of forward movement, relative ratios, comparative numbers of a 200-day moving average, strength index, volatility targets, moving average convergence divergence (MACD), fundamental value, etc. The subsequent sections in this chapter deal with each of the relevant front-end tools recommended for practitioners.

8.3.1 Moving Average Convergence Divergence (MACD)

Moving averages are simple moving averages, the average of days of closing prices or exponential moving average (EMA). It is calculated as:

$$\mathrm{EMA_T} = P_T\left\{1 + \frac{K}{1+N}\right\} + \mathrm{EMA_{T-1}}\left\{1 - \frac{K}{1+N}\right\} \tag{8.1}$$

Table 8.1 Trading journal entry record

Serial number trade #	Date	Entry time	Symbol	Long or short	Entry price	Stop-Loss	QTY	Risk [Q x (EP – EP)]	Exit price	Exit date	Exit time
1	1/17/2006	10.2	AAPL	L	84.8	80	900	450	85.3	1/19/2006	15.22
2	10/1/2011	10.2	ACC	L	420.00	410.00	15	150	430.00	10/1/2011	15.22
3	12/15/2011	10.2	ABA	L	920.00	910.00	15	150	930.00	12/15/2011	15.22
4	1/1/2012	10.2	AJA	L	425.00	415.00	15	150	435.00	1/1/2012	15.22
5	2/18/2012	10.2	CIPLA	L	400.00	390.00	15	150	410.00	2/18/2012	15.22
6	4/1/2012	10.2	AXIS	L	410.00	400.00	15	150	420.00	4/1/2012	15.22
7	7/1/2012	10.2	YES	L	25.00	15.00	15	150	35.00	7/1/2012	15.22
8	10/1/2012	10.2	ICICI	L	226.00	216.00	15	150	236.00	10/1/2012	15.22
9	10/17/2012	10.2	XYZ	L	545.00	535.00	15	150	555.00	10/17/2012	15.22
10	10/30/2012	10.2	ANA	L	120.00	110.00	15	150	130.00	10/30/2012	15.22

where N is the number of days, P_{T-1} is the previous day's closing price, EMA_{T-1} is the exponential moving average for the previous day, and K is the smoothing factor that equals 2. The EMA is used against the SMA (simple moving average) as the EMA includes recent price points. Table 8.2 shows EMA calculation from Daily price data.

MACD is the shift in the direction of a given trend in prices. The MACD is calculated by the difference between the 26-period Exponential moving averages (EMAs) derived from the 12-period EMA for the same set of prices. Later, a 9-day EMA of the MACD is used to forecast the MACD to gauge the direction of the trend.

The MACD is expressed as:

$$MACD = \mathrm{EMA}_{12} - \mathrm{EMA}_{26} \tag{8.2}$$

Table 8.3 shows the data for calculating MACD alongside the date and the daily closing prices of ABC stock. EMA for 12 periods is the average of closing prices of the first successive 12 days closing prices. Using the equation, the 12-day EMA is given as:

$EMA_t = P_t \times K + EMA_{t-1} \times (1 - K) = 11.75 \text{ x } (2/(12 + 1) + (10.79 \text{ x } (1 - (2/(12 + 1)) = 10.94$.

The closing price is 11.75.

Table 8.2 EMA calculation from daily data

Smoothing factor	2				
Day	Price	Previous price	SMA	Exponential moving Average - previous	Exponential moving Average
1	155				
2	200	155			
3	158	200	171.0	171.0	171.0
4	130	158	162.7	171.0	171.0
5	100	130	129.3	171.0	171.0
6	160	100	130.3	171.0	171.0
7	170	160	143.3	171.0	171.0
8	100	170	143.3	171.0	171.0
9	150	100	140.0	171.0	171.0
10	200	150	150.0	171.0	171.0
11	280	200	210.0	171.0	171.0
12	210	280	230.0	171.0	171.0
13	175	210	221.7	171.0	171.0
14	155	175	180.0	171.0	171.0
15	160	155	180.0	171.0	171.0

Table 8.3 Forward trigger 9 days MACD using daily closing price

No	Date	Closing price	12-Period EMA	26-Period EMA	MACD	Forward Trigger 9 Days
1	1/5/2019	101	100.154	100	0.153846	
2	2/5/2019	102	100.308	100	0.307692	
3	3/5/2019	103	100.462	100	0.461538	
4	3/6/2019	105	100.769	100	0.769231	
5	3/7/2019	101	100.154	100	0.153846	
6	3/8/2019	101	100.154	100	0.153846	
7	3/9/2019	102	100.308	100	0.307692	
8	3/10/2019	102	100.308	100	0.307692	
9	3/11/2019	103	100.462	100	0.461538	
10	3/12/2019	104	100.615	100	0.615385	
11	3/13/2019	104	100.615	100	0.615385	
12	3/14/2019	105	100.769	100	0.769231	
13	3/15/2019	106	100.923	100	0.923077	
14	3/16/2019	109	101.385	100	1.384615	
15	3/17/2019	101	100.154	100	0.153846	
16	3/18/2019	102	100.308	100	0.307692	
17	3/19/2019	103	100.462	100	0.461538	
18	3/20/2019	105	100.769	100	0.769231	
19	3/21/2019	101	100.154	100	0.153846	
20	3/22/2019	101	100.154	100	0.15	
21	3/23/2019	102	100.308	100	0.31	
22	3/24/2019	102	100.308	100	0.31	
23	3/25/2019	103	100.462	100	0.46	
24	3/26/2019	104	100.615	100	0.62	
25	3/27/2019	104	100.615	100	0.62	
26	3/28/2019	105	100.769	100	0.77	
27	3/29/2019	106	100.923	100	0.92	0.462
28	3/30/2019	109	101.385	100	1.38	0.462
29	3/31/2019	101	100.154	100	0.15	0.462
30	4/1/2019	102	100.308	100	0.31	0.462
31	4/2/2019	103	100.462	100	0.46	0.462
32	4/3/2019	105	100.769	100	0.77	0.462
33	4/4/2019	101	100.154	100	0.15	0.462
34	4/5/2019	101	100.154	100	0.15	0.462
35	4/6/2019	102	100.308	100	0.31	0.462
36	4/7/2019	102	100.308	100	0.31	0.462
37	4/8/2019	103	100.462	100	0.46	0.462
38	4/9/2019	104	100.615	100	0.62	0.462
39	4/10/2019	104	100.615	100	0.62	0.462
		k	0.15			

The closing price is 11.75. The 26-day EMA is calculated for the successive period of 26 days. The averages of the closing prices for 26 days, where the 26-period EMA are as follows:

EMA = $P_t \times K + EMA_{t-1}$ (1 − K) = (12.47 x (2/26 + 1) + (11.41 × (1–(2 / (26 + 1))) = 11.46.

The closing price is 12.47.

The 12-period EMA is subtracted from the 26-day EMA. The fast line is the difference between a running line and a slow line. The forward values for the base (signal) line are the 9-day EMA of the MACD. The EMA for 9 Days of the MACD is the average number of the first successive 9 MACD values.

The 9-day EMA is calculated by using the values of the MACD.

$$\mathrm{EMA}_{9\mathrm{day}} = \frac{0.71\mathrm{x}\frac{2}{9+1}}{0.66\mathrm{x}(1-\frac{2}{9+1})} = 0.67$$

The slow line in MACD is the 9-day moving average. The fast line in MACD is calculated from the difference in the 26-day exponential moving average (EMA) and the 12-day exponential moving average (EMA), respectively. When the MACD falls lower than the trigger of 9 days forward MACD, it is a sell. This is a bearish sign. Alternatively, when the MACD rises above the trigger of 9 days forward MACD, it is a buy. As shown in the table, the fast line crosses the slow line at row 25. Similarly, at row 23, the slow line crosses the fast line. Table 8.4 shows the calculation of v-MACD. For the jerky day on 03/29/2019, row 27, the v-MACD turns (<0) and is given as:

= SMA(14) –v-MA = 102.75–104.00 = −1.25.

8.3.2 Volatility Adjusted MACD

It implies making adjustments to moving averages against a changing volatility level. The practitioner could use the standard definition of ATR $\{P_{MAX} = P_{MIN}\}$ the average True Range or volatility ratio ($Ratio = \frac{\{P_{MAX}=P_{MIN}\}}{AVERAGE\{P_{MAX}=P_{MIN}\}}$) to adjust the MACD. Usually, a volatility ratio above 0.5 is used to identify reversals or upheavals. V-MACD (Volatility adjusted MACD) is calculated as the difference between two moving averages. The V-MACD uses Volatility adjusted Moving Average (V-MA) as Slow MA and Simple Moving Average (SMA) as Fast MA:

$$V_{\mathrm{MACD}} = \mathrm{SMA} \quad V_{MA} \tag{8.3}$$

where,

$$V_{\mathrm{MA}} = \mathrm{EMA}_{26}\{\mathrm{IF\ Ratio}_9 \geq 0.8, \mathrm{else}\} = \mathrm{EMA}_{12} \tag{8.4}$$

Table 8.4 Forward trigger 9 days V-MACD using daily closing price

No	Date	Closing Price	EMA (12)	EMA (26)	V-MACD	SMA (14)	V-MA		Ratio (9)
1	1/5/2019	101	100.153846	100					
2	2/5/2019	102	100.307692	100					
3	3/5/2019	103	100.461538	100					
4	3/6/2019	105	100.769231	100					
5	3/7/2019	101	100.153846	100					
6	3/8/2019	101	100.153846	100					
7	3/9/2019	102	100.307692	100					
8	3/10/2019	102	100.307692	100				4	0.80
9	3/11/2019	103	100.461538	100				4	0.73
10	3/12/2019	104	100.615385	100				4	0.68
11	3/13/2019	104	100.615385	100				4	0.63
12	3/14/2019	105	100.769231	100				4	0.59
13	3/15/2019	106	100.923077	100		102.75	100.92	4	0.55
14	3/16/2019	109	101.384615	100		103.17	101.38	5	0.65
15	3/17/2019	101	100.153846	100	3.75	103.75	100.00	8	1.00
16	3/18/2019	102	100.307692	100	3.58	103.58	100.00	8	1.00
17	3/19/2019	103	100.461538	100	3.33	103.33	100.00	8	1.00
18	3/20/2019	105	100.769231	100	3.50	103.50	100.00	8	1.00
19	3/21/2019	101	100.153846	100	3.83	103.83	100.00	8	1.00
20	3/22/2019	101	100.153846	100	3.75	103.75	100.00	8	0.81
21	3/23/2019	102	100.307692	100	3.36	103.67	100.31	8	0.68
22	3/24/2019	102	100.307692	100	3.28	103.58	100.31	8	0.59
23	3/25/2019	103	100.461538	100	2.96	103.42	100.46	8	0.51
24	3/26/2019	104	100.615385	100	2.72	103.33	100.62	8	0.46
25	3/27/2019	104	100.615385	100	2.63	103.25	100.62	8	0.41
26	3/28/2019	105	100.769231	100	2.31	103.08	100.77	8	0.38
27	3/29/2019	126	104.004562	100	−1.25	102.75	104.00	8	0.35
28	3/30/2019	109	101.384615	100	4.83	104.83	100.00	25	0.96
29	3/31/2019	101	100.153846	100	5.42	105.42	100.00	25	0.92
30	4/1/2019	102	100.307692	100	5.25	105.25	100.00	25	0.88
31	4/2/2019	103	100.461538	100	5.00	105.00	100.00	25	0.85
32	4/3/2019	105	100.769231	100	5.17	105.17	100.00	25	0.83
33	4/4/2019	101	100.153846	100	5.50	105.50	100.00	25	0.81
34	4/5/2019	101	100.153846	100	5.26	105.42	100.15	25	0.79
35	4/6/2019	136	105.538462	100	−0.21	105.33	105.54	25	0.76
36	4/7/2019	102	100.307692	100	8.08	108.08	100.00	35	1.00
37	4/8/2019	103	100.461538	100	7.92	107.92	100.00	35	1.00
38	4/9/2019	104	100.615385	100	7.83	107.83	100.00	35	1.00
39	4/10/2019	104	100.615385	100	7.75	107.75	100.00	35	1.00
		k	0.15						

Volatility adjusted MA uses the inputs of the volatility ratio calculated above. The Volatility adjusted MA, in a period of low volatility, is equal to MACD. V-MACD behaves as MACD in periods of high volatility. Positive V-MACD suggests bullish sentiment and negative V-MACD readings indicate bearish sentiment. One would buy when V-MACD becomes positive and sell when V-MACD drops into negative.

It would mean Fast moving average is a 12-day Simple Moving Average (SMA). Slow Moving Average is Volatility Adjusted Moving Average, which moves exactly as SMA (26) for as long as 9-day Absolute ATR is below 0.8%. When 9-day Absolute ATR crosses above 0.8%, the 26-day Moving Average will be adjusted to volatility (Absolute ATR) level. Therefore, buy when V-MACD becomes positive. Sell when V-MACD drops to attain negative.

8.3.3 200-Day Moving Average

The 200-day moving average is the simple average of consecutively traded prices. When the price on a day rises above the moving average, it's a buy signal, and when the price falls below the moving average, it is a sell signal.

The formula is:

$$P_{200} = \frac{P_1 + P_2 + P_3 \ldots\ldots + P_{200}}{200} \tag{8.5}$$

where p_t is the daily closing price.

The 200-day moving average is a long-term moving average. When the price action breaks the 200-day upwards, it gives a long signal. When the price action breaks the 200-day SMA downwards, it creates a short signal. The other alternatives to 200 days are the 50-day moving average or the 10-day moving average, etc. The more volatile is the stock, the longer is the period for calculating the moving average. Table 8.5 shows the 200-Day Moving Average.

8.3.4 Support and Resistance

When the price increases during an upward trend and the open interest also rises, it is undertaken as new buying and bullish. If the price rises and the open interest declines, the short-sellers have covered positions and money has started leaving the market and bearish. A rise in price trend and rise in open interest are buy signals. Falling open interest is the end of a trend. When prices are in a downtrend, and open interest (OI) is rising, it could rise further. If the open interest is reducing and the prices are in a downward trend, the nearest bottom could reach sooner. Table 8.6 shows Basic Rules for Volume and Open Interest. Table 8.7 shows the Support and resistance from Open Interest volume.

Table 8.5 200-day moving average

Date	Closing price	Moving average	Trigger
10/5/2015	113.2	108.32	1
10/6/2015	110.5	108.36	1
10/7/2015	108.3	108.15	1
10/8/2015	108.2	108.67	0
10/9/2015	111.7	108.63	1
10/12/2015	107.3	108.32	0
10/13/2015	104.9	108.36	0
10/14/2015	104.8	108.15	0
10/15/2015	104.7	108.67	0
10/16/2015	104.5	108.63	0
10/19/2015	106.2	104.39	1
10/20/2015	107.6	104.37	1
10/21/2015	108.8	104.32	1
10/22/2015	108.8	104.25	1
10/23/2015	110.0	104.18	1
10/26/2015	110.2	104.09	1
10/27/2015	109.9	104.00	1
10/28/2015	110.3	103.91	1
10/29/2015	109.5	103.80	1
10/30/2015	108.5	103.71	1
11/2/2015	108.1	103.63	1
11/3/2015	109.6	103.55	1
11/4/2015	108.7	103.45	1
11/5/2015	107.3	103.36	1
11/6/2015	108.8	103.29	1
11/9/2015	108.5	103.19	1
11/10/2015	105.4	103.09	1
11/11/2015	105.4	103.04	1
11/12/2015	105.4	103.00	1
11/13/2015	105.1	102.95	1
11/16/2015	103.0	102.91	1
11/17/2015	101.2	102.91	0
11/18/2015	97.0	102.94	0
11/19/2015	99.8	103.07	0
11/20/2015	100.2	103.14	0
11/23/2015	100.4	103.21	0
11/24/2015	99.1	103.27	0
11/25/2015	99.1	103.37	0
11/26/2015	100.2	103.47	0
11/27/2015	101.6	103.55	0
11/30/2015	103.8	103.60	1
12/1/2015	102.9	103.59	0

(continued)

Table 8.5 (continued)

Date	Closing price	Moving average	Trigger
12/2/2015	101.1	103.61	0
12/3/2015	100.8	103.68	0
12/4/2015	99.9	103.76	0
12/7/2015	99.6	103.87	0
12/8/2015	99.4	103.99	0
12/9/2015	97.8	104.13	0
12/10/2015	99.7	104.33	0
12/11/2015	100.3	104.48	0
12/14/2015	102.0	104.62	0
12/15/2015	102.8	104.71	0
12/16/2015	104.6	104.78	0
12/17/2015	105.7	104.78	1
12/18/2015	103.3	104.75	0
12/21/2015	105.3	104.81	1
12/22/2015	103.4	104.78	0
12/23/2015	105.1	104.85	1
12/24/2015	104.6	104.83	0
12/25/2015	104.6	104.84	0
12/28/2015	105.5	104.85	1
12/29/2015	105.5	104.82	1
12/30/2015	103.6	104.79	0
12/31/2015	105.5	104.85	1
1/1/2016	105.5	104.81	1
1/4/2016	102.9	104.76	0
1/5/2016	102.4	104.90	0
1/6/2016	101.9	105.09	0
1/7/2016	100.1	105.35	0
1/8/2016	101.3	105.83	0
1/11/2016	100.6	106.28	0
1/12/2016	100.0	106.92	0
1/13/2016	103.3	107.78	0
1/14/2016	108.3	108.41	0
1/15/2016	109.0	108.43	1
1/18/2016	108.2	108.32	0
1/19/2016	109.0	108.36	1
1/20/2016	107.1	108.15	0
1/21/2016	108.7	108.67	1
1/22/2016	108.6	108.63	0

Table 8.6 Basic rules for volume and open interest

No	Price	Volume	Open interest	Observation
1	Up	UP	UP	Long buildup
2	UP	DOWN	DOWN	Short covering
3	Down	UP	UP	Short buildup
4	Down	DOWN	DOWN	Long buildup

8.3.5 RSI

RSI is a non-parametric count of the ratio between rises and falls. RSI is presented to vary between 0 and 100 (0–100). For example, when the RSI is greater than 70, the stock is bought over, and if the RSI is lower than 30, the stock is sold over.

The gain or loss for a single period is based on the change of the current period's closing price less the previous period's closing price. The average stock price gain is the sum of the gains divided by 14. The average stock price loss is the sum of the losses divided by 14 (As shown in Table 8.8).

$$\mathrm{RSI} = 100 - \frac{100}{1 - RS} \tag{8.6}$$

$$\mathrm{RS} = \frac{\text{Sum of gains during higher closing prices/14}}{\text{Sum of losses during a lower closing price/14}} \tag{8.7}$$

8.3.6 EWMA Volatility Strategy

The exponential-weighted moving average volatility of a portfolio arrives at the target volatility of a portfolio. In this approach, a smoothing factor (=0.94 or lower) is applied to past returns to calculate the volatility.

The exponential-weighted moving averages are calculated using the formulae as follows:

$$\rho_t^2 = 261 \times \sum\nolimits_0^t (1 - \delta)\delta^i (r_{t-i-1} - \dot{r})^2 \tag{8.8}$$

where ρ_t^2 is the exponential-weighted volatility,

δ is the factor that determines the weight on past daily volatility,

$$w_i = (1 - \delta)\delta^i \tag{8.9}$$

where i is the number of days since the earliest day of trading returns. The smoothing factor δ *is equal to* 0.94.

Table 8.7 Support and resistance from open interest volume

Calls										Puts							
Scenario	Interpretation	OI change	Price change	Call volume	Call Net OI	Call change in OI	Call Price change	Call LTP	Strike	Put LTP	Put price change	Put change in OI	Put net OI	Put volume	Price change	OI change	Interpretation
1	Short buildup	UP	DOWN	119	105	37	-489.85	803.45	8500	18.55	14.15	6273	14,302	116,151	UP	UP	Long buildup
2	Short buildup	UP	DOWN	5	4	1	–453	757	8600	25.05	19	1021	3826	60,665	UP	UP	Long buildup
3	Short buildup	UP	DOWN	37	14	9	–512.1	597.9	8700	33.55	26.3	1933	5693	80,164	UP	UP	Long buildup
4	Short buildup	UP	DOWN	85	55	18	–481	535	8800	44.85	36.55	3918	7760	118,211	UP	UP	Long buildup
5	Short buildup	UP	DOWN	70	42	17	–435	455	8900	59.5	49.3	4222	6149	104,833	UP	UP	Long buildup
6	Short buildup	UP	DOWN	3142	1201	433	–457.6	365.1	9000	79	66.75	5703	15,835	307,611	UP	UP	Long buildup
7	Short buildup	UP	DOWN	1532	651	256	–438.3	295.2	9100	104.25	89	2582	6049	150,186	UP	UP	Long buildup
8	Short buildup	UP	DOWN	10,420	1452	1150	–407	220.15	9200	136.9	117.55	2605	9729	206,103	UP	UP	Long buildup
9	Short buildup	UP	DOWN	32,685	4244	4432	–362.7	163	9300	176.2	150.7	2154	9701	264,338	UP	UP	Long buildup
10	Short buildup	UP	DOWN	105,520	9956	5195	–327.4	114.8	9400	227.9	192.45	63	6749	183,627	UP	UP	Long buildup
11	Short buildup	UP	DOWN	142,101	15,221	13,560	–276.65	78.95	9500	292.45	242.75	–6491	7086	129,809	UP	DOWN	Short covering
12	Short buildup	UP	DOWN	120,112	11,302	10,235	–221.85	52	9600	364.05	293.65	–6434	3006	41,154	UP	DOWN	Short covering
13	Short buildup	UP	DOWN	106,315	10,820	6001	–172.3	33.3	9700	445.05	345.25	–7667	2712	32,561	UP	DOWN	Short covering

(continued)

Table 8.7 (continued)

Calls										Puts							
Scenario	Interpretation	OI change	Price change	Call volume	Call Net OI	Call change in OI	Call Price change	Call LTP	Strike	Put LTP	Put price change	Put change in OI	Put net OI	Put volume	Price change	OI change	Interpretation
14	Short buildup	UP	DOWN	111,650	17,512	11,300	–125.3	21.5	9800	532	390.7	–4840	3226	22,643	UP	DOWN	Short covering
15	Short buildup	UP	DOWN	10,715	9520	1432	–86.8	14	9900	625	430.9	–1313	733	4673	UP	DOWN	Short covering
16	Short buildup	UP	DOWN	124,452	21,332	17,236	–56.45	9.5	10,000	716.5	459.65	–726	635	3289	UP	DOWN	Short covering
17	Short buildup	UP	DOWN	34,113	4650	1285	–35.2	5.95	10,100	816.45	484.1	–20	72	79	UP	DOWN	Short covering

Table 8.8 RSI detection and strategy

Date	Closing price	Change in price	Gain	Loss	14-Day average gain	14-Day average loss	RS	14-Day RSI
	1	2	3	4	5	6	7	8
10/5/2015	113.2	–2.7	0.0	2.7				
10/6/2015	110.5	–2.1	0.0	2.1				
10/7/2015	108.3	–0.1	0.0	0.1				
10/8/2015	108.2	3.5	3.5	0.0				
10/9/2015	111.7	–4.5	0.0	4.5				
10/12/2015	107.3	–2.3	0.0	2.3				
10/13/2015	104.9	–0.2	0.0	0.2				
10/14/2015	104.8	0.0	0.0	0.0				
10/15/2015	104.7	–0.2	0.0	0.2				
10/16/2015	104.5	1.7	1.7	0.0				
10/19/2015	106.2	1.4	1.4	0.0				
10/20/2015	107.6	1.2	1.2	0.0				
10/21/2015	108.8	0.0	0.0	0.0				
10/22/2015	108.8	1.2	1.2	0.0	0.6	0.9	0.7	42.5
10/23/2015	110.0	0.2	0.2	0.0	0.6	0.8	0.8	43.1
10/26/2015	110.2	–0.3	0.0	0.3	0.6	0.8	0.7	42.5
10/27/2015	109.9	0.4	0.4	0.0	0.6	0.7	0.8	43.7
10/28/2015	110.3	–0.8	0.0	0.8	0.5	0.7	0.7	41.7
10/29/2015	109.5	–1.0	0.0	1.0	0.5	0.7	0.6	39.3
10/30/2015	108.5	–0.4	0.0	0.4	0.4	0.7	0.6	38.3
11/2/2015	108.1	1.4	1.4	0.0	0.5	0.7	0.8	43.6
11/3/2015	109.6	–0.9	0.0	0.9	0.5	0.7	0.7	41.3
11/4/2015	108.7	–1.4	0.0	1.4	0.4	0.7	0.6	37.8
11/5/2015	107.3	1.5	1.5	0.0	0.5	0.7	0.8	43.5
11/6/2015	108.8	–0.4	0.0	0.4	0.5	0.7	0.7	42.5
11/9/2015	108.5	–3.1	0.0	3.1	0.5	0.8	0.5	35.2
11/10/2015	105.4	0.0	0.0	0.0	0.4	0.8	0.5	35.2
11/11/2015	105.4	0.0	0.0	0.0	0.4	0.7	0.5	35.2
11/12/2015	105.4	–0.3	0.0	0.3	0.4	0.7	0.5	34.5
11/13/2015	105.1	–2.1	0.0	2.1	0.3	0.8	0.4	30.0
11/16/2015	103.0	–1.9	0.0	1.9	0.3	0.9	0.4	26.5
11/17/2015	101.2	–4.2	0.0	4.2	0.3	1.1	0.3	20.9
11/18/2015	97.0	2.8	2.8	0.0	0.5	1.0	0.5	31.4
11/19/2015	99.8	0.5	0.5	0.0	0.5	0.9	0.5	33.0
11/20/2015	100.2	0.1	0.1	0.0	0.4	0.9	0.5	33.5
11/23/2015	100.4	–1.3	0.0	1.3	0.4	0.9	0.5	31.2
11/24/2015	99.1	0.0	0.0	0.0	0.4	0.8	0.5	31.2
11/25/2015	99.1	1.1	1.1	0.0	0.4	0.8	0.6	35.8

(continued)

Table 8.8 (continued)

Date	Closing price	Change in price	Gain	Loss	14-Day average gain	14-Day average loss	RS	14-Day RSI
11/26/2015	100.2	1.4	1.4	0.0	0.5	0.7	0.7	41.0
11/27/2015	101.6	2.2	2.2	0.0	0.6	0.7	0.9	48.0
11/30/2015	103.8	−0.9	0.0	0.9	0.6	0.7	0.8	45.6
12/1/2015	102.9	−1.8	0.0	1.8	0.5	0.8	0.7	41.0
12/2/2015	101.1	−0.3	0.0	0.3	0.5	0.7	0.7	40.3
12/3/2015	100.8	−0.9	0.0	0.9	0.5	0.7	0.6	38.3
12/4/2015	99.9	−0.3	0.0	0.3	0.4	0.7	0.6	37.6
12/7/2015	99.6	−0.2	0.0	0.2	0.4	0.7	0.6	37.1
12/8/2015	99.4	−1.6	0.0	1.6	0.4	0.7	0.5	33.3
12/9/2015	97.8	1.9	1.9	0.0	0.5	0.7	0.7	41.1
12/10/2015	99.7	0.5	0.5	0.0	0.5	0.6	0.8	43.1
12/11/2015	100.3	1.7	1.7	0.0	0.6	0.6	1.0	48.9
12/14/2015	102.0	0.8	0.8	0.0	0.6	0.6	1.1	51.5
12/15/2015	102.8	1.8	1.8	0.0	0.7	0.5	1.3	56.9
12/16/2015	104.6	1.1	1.1	0.0	0.7	0.5	1.5	59.8
12/17/2015	105.7	−2.4	0.0	2.4	0.7	0.6	1.1	51.7
12/18/2015	103.3	2.0	2.0	0.0	0.8	0.6	1.3	57.0
12/21/2015	105.3	−2.0	0.0	2.0	0.7	0.7	1.0	51.2
12/22/2015	103.4	1.7	1.7	0.0	0.8	0.6	1.2	55.4
12/23/2015	105.1	−0.5	0.0	0.5	0.7	0.6	1.2	54.1
12/24/2015	104.6	0.0	0.0	0.0	0.7	0.6	1.2	54.1
12/25/2015	104.6	0.8	0.8	0.0	0.7	0.5	1.3	56.4
12/28/2015	105.5	0.0	0.0	0.0	0.6	0.5	1.3	56.4
12/29/2015	105.5	−1.8	0.0	1.8	0.6	0.6	1.0	50.0
12/30/2015	103.6	1.9	1.9	0.0	0.7	0.5	1.3	55.6
12/31/2015	105.5	0.0	0.0	0.0	0.6	0.5	1.2	55.6
1/1/2016	105.5	−2.6	0.0	2.6	0.6	0.7	0.9	47.1
1/4/2016	102.9	−0.5	0.0	0.5	0.5	0.6	0.8	45.8
1/5/2016	102.4	−0.5	0.0	0.5	0.5	0.6	0.8	44.4
1/6/2016	101.9	−1.9	0.0	1.9	0.5	0.7	0.7	39.5
1/7/2016	100.1	1.3	1.3	0.0	0.5	0.7	0.8	44.0
1/8/2016	101.3	−0.8	0.0	0.8	0.5	0.7	0.7	41.9
1/11/2016	100.6	−0.6	0.0	0.6	0.5	0.7	0.7	40.4
1/12/2016	100.0	3.3	3.3	0.0	0.7	0.6	1.1	51.5
1/13/2016	103.3	5.0	5.0	0.0	1.0	0.6	1.7	62.7
1/14/2016	108.3	0.7	0.7	0.0	0.9	0.5	1.8	63.9
1/15/2016	109.0	−0.8	0.0	0.8	0.9	0.6	1.6	61.4
1/18/2016	108.2	0.8	0.8	0.0	0.9	0.5	1.7	62.9
1/19/2016	109.0	−1.8	0.0	1.8	0.8	0.6	1.3	57.1

(continued)

Table 8.8 (continued)

Date	Closing price	Change in price	Gain	Loss	14-Day average gain	14-Day average loss	RS	14-Day RSI
1/20/2016	107.1	1.6	1.6	0.0	0.9	0.6	1.5	60.5
1/21/2016	108.7	–0.1	0.0	0.1	0.8	0.5	1.5	60.2
1/22/2016	108.6	–108.6	0.0	108.6	0.7	8.3	0.1	8.3

As shown in the table below, the target volatility of a portfolio may rise or fall depending on the pattern of past volatility. A smoothing factor of 0.94 using 30 days of historical data gives the daily volatility levels for 6 days, starting from 3.92% to 7.99%. A small variation in price ensures wider displayed ranges of EWMA volatility (Table 8.9).

The EWMA for Day 2 on 3/25/2018 is calculated as:

= SUMPRODUCT (squared return × weight) × 261 = 14.23.

In a volatility strategy, the manager may alter the leverage to match the target volatility of the portfolio. If the volatility falls, the leverage is enhanced. The manager may alter each stock's individual weights when the portfolio's EWMA volatility is higher than his target level.

8.3.7 Equal Volatility Weighting

Instead of allocating budget on the market size of the stock, the equal volatility weighting gives weights based on the degree of riskiness. In particular, it assigns lower weights to higher risks, and it assigns more weight against lower total volatility of the asset. The result of the approach makes the contribution of every asset to overall portfolio volatility equal.

The equal volatility formula is given as:

$$W_i = \frac{1}{\sigma_i} / \{ \frac{1}{\sigma_1} + \frac{1}{\sigma_2} + \frac{1}{\sigma_3} \dots + \frac{1}{\sigma_j} + \dots \} \tag{8.4}$$

Table 8.10 shows the computation of equal volatility–weighted portfolio. The changes to weights are shown alongside in the next period.

The first-period weight for Stock D is calculated as:

$$= \frac{1}{7} / \{ \frac{1}{5} + \frac{1}{10} + \frac{1}{10} + \frac{1}{7} + \frac{1}{9} = 21.84$$

The limitation is when the volatility of a security is unusually low, larger weights are assigned to it. When the volatility falls back to a previous normal level, the weights are too large to be reviewed. It does not take into consideration correlations.

Table 8.9 EWMA volatility

Day	Price	Daily return	Squared return	EWMA	Daily volatility	Lambda	0.94
12/31/2018	101	–		–	–	Days count	Weight
1/1/2018	102	0.99%	0.00980%	–	–	30	0.938%
1/2/2018	103	0.98%	0.00961%	–	–	29	0.997%
1/3/2018	105	1.94%	0.03770%	–	–	28	1.061%
1/6/2018	101	–3.81%	0.14512%	–	–	27	1.129%
1/7/2018	101	0.00%	0.00000%	–	–	26	1.201%
1/8/2018	102	0.99%	0.00980%	–	–	25	1.277%
1/9/2018	102	0.00%	0.00000%	–	–	24	1.359%
1/10/2018	103	0.98%	0.00961%	–	–	23	1.446%
1/13/2018	104	0.97%	0.00943%	–	–	22	1.538%
1/14/2018	104	0.00%	0.00000%	–	–	21	1.636%
1/15/2018	105	0.96%	0.00925%	–	–	20	1.741%
1/16/2018	106	0.95%	0.00907%	–	–	19	1.852%
1/17/2018	109	2.83%	0.08010%	–	–	18	1.970%
1/20/2018	101	–7.34%	0.53868%	–	–	17	2.096%
1/21/2018	102	0.99%	0.00980%	–	–	16	2.229%
1/22/2018	103	0.98%	0.00961%	–	–	15	2.372%
1/23/2018	105	1.94%	0.03770%	–	–	14	2.523%
1/24/2018	101	–3.81%	0.14512%	–	–	13	2.684%
1/27/2018	101	0.00%	0.00000%	–	–	12	2.856%
1/28/2018	102	0.99%	0.00980%	–	–	11	3.038%
1/29/2018	102	0.00%	0.00000%	–	–	10	3.232%
1/30/2018	103	0.98%	0.00961%	–	–	9	3.438%
1/31/2018	104	0.97%	0.00943%	–	–	8	3.657%
2/3/2018	104	0.00%	0.00000%	–	–	7	3.891%
2/4/2018	105	0.96%	0.00925%	–	–	6	4.139%
2/5/2018	106	0.95%	0.00907%	–	–	5	4.403%
2/6/2018	109	2.83%	0.08010%	–	–	4	4.684%
2/7/2018	101	–7.34%	0.53868%	–	–	3	4.984%
2/10/2018	102	0.99%	0.00980%	–	–	2	5.302%
2/11/2018	103	0.98%	0.00961%	–	–	1	5.640%
3/25/2018	105	1.94%	0.03770%	14.23%	37.73%		
3/26/2018	101	–3.81%	0.14512%	13.91%	37.30%		
3/27/2018	101	0.00%	0.00000%	15.19%	38.98%		0.79
3/30/2018	102	0.99%	0.00980%	14.19%	37.67%		
3/31/2018	102	0.00%	0.00000%	13.15%	36.27%		
4/1/2018	103	0.98%	0.00961%	12.36%	35.16%		

Table 8.10 Equal-weighted portfolio

Stock	Period 1	Period 2	Period 1	Period 2	
	Volatility (%)	Volatility (%)	Weight (%)	Weight (%)	Change
A	5.00	5.00	30.58	34.69	Rise
B	10.00	7.00	15.29	24.78	Rise
C	10.00	9.00	15.29	19.27	Rise
D	7.00	10.00	21.84	17.35	Fall
E	9.00	10.00	16.99	17.35	Rise

8.3.8 Fair Value Measurement

One may attempt to derive the fundamental value using the Ben Graham Formula. A conservative estimate of the constant growth number is desirable since no firm can grow forever at a rate higher than the economy's growth rate in which it operates. It is critical to determine the loss of the competitive advantage and estimate the number of years that companies can maintain their extraordinary returns. Table 8.11 shows the Fundamental Value using the Ben Graham Formula.

8.3.9 Covid-19 Shocks

The portfolio impact of COVID-19 and simulated stock prices is a challenge. The post-simulated earnings are arrived at with the sole aim of distinguishing the price impact due to staggered earnings versus the long spell of turbulence in the market. Investors react negatively during the period of shock, whereas the growth rates appear larger than the actual period of the COVID spell. The COVID-19 has resulted in gains to sectors such as Pharma, Insurance, Basic Staple Consumers, Insurance, Telecom, and chemicals. COVID-19 has caused losses to Travel and Hospitality, Health Care, Real Estate, Banking and Financial services, etc. Table 8.12 shows two standard sectors of 1. Pharmaceutical and 2. Reality sectors, respectively. The interesting aspect is to arrive at the contrasting evidence of

Table 8.11 Fundamental value: The Ben Graham formula

The Ben Graham Formula	
Company Name	ABC CEMENT LTD
Year Ended	Oct/75
Avg 5-Yr Net Profit (Rs Crore)	163.4
PE Ratio at 0% Growth	8.5
Long-Term Growth Rate	5.9
The Ben Graham Value (Rs Crore)	3,325
Current Market Cap (Rs Crore)	---
Ben Graham's Original Formula: Value = EPS x (8.5 + 2G)	

Table 8.12 COVID-19 impact on reality and pharma sector

Reality	Fall in index (%)		Pharma	Reality	Standard deviation of index 2020		Pharma
Month	Mean	Month	Mean	Month	Mean	Month	Mean
July	–39%	July	20%	July	201.0	July	10,113.7
June	–37%	June	19%	June	204.0	June	10,036.2
May	–54%	May	7%	May	170.1	May	9324.6
April	–49%	April	–6%	April	181.8	April	8926.8
@ all data for 2020							
July	–78	July	2032	July	279.2	July	8081.8
June	–73	June	1929	June	276.9	June	8107.5
May	–94	May	611	May	264.1	May	8713.6
April	–89	April	–435	April	270.6	April	9361.9

@ all data for 2019
SourceAuthor's calculation using NSE data 2020

COVID impact between the two sectors. We present the delta to the monthly mean indices and monthly standard deviations between 2019 and 2020 for April, May, June, and July, respectively.

The COVID-19 has severely impacted the Reality sector when the indices have fallen from 73 to 94 points ranging between 37 to 49% drop in indices. However, the volatility of the Reality sector has also dropped from levels of 264 to 279 points to about 170 to 204 points, in 2020. The pharma sector has gained positively, from –435 points to 2032 points that equaled –6 to 20% in 2020. The volatility of Pharma sector stocks has risen from 8107 to 9361 points to levels of 8926 to 10,113 points in 2020. The inference for the investor is to reduce the weights on Reality stocks and enhance the weights on Pharma stocks.

8.4 Summary

The sole focus of this chapter was to provide a variety of tools for learning and applying at the trading desk. It discussed important aspects of the trading desk, data, software, analysis outputs, inferences, and validation in practice. The true nature of its application will indeed vary from desk to desk, individuals to individuals in real life. The purpose of learning these techniques is not to become an expert trader, but to understand that a strategy that works today could fail tomorrow. It may fail tomorrow because the known strategy, however, tested for umpteen times, cannot make for the randomness in its spell.

References

Azoff, E.M. Neural Network Time Series Forecasting of Financial Markets John Wiley and Sons Ltd, 1994.

Christoffersen, P.F., and F.X. Diebold. 2006. Financial asset returns, direction-of-change forecasting, and volatility dynamics. *Management Science* 52 (8): 1273–1287.

Graham, B. The Intelligent Investor HarperCollins; Rev Ed edition, 2003.

Lo, A.W. and Mackinlay, A.C. A Non-Random Walk Down Wall Street 5th, Ed. Princeton University Press, 2002.

Murphy John, Technical Analysis of the Financial Markets: A Comprehensive Guide to Trading Methods and Applications (New York Institute of Finance), 1999.

Appendix
Data and Tables

Table A1 Realized volatility of ABC Stock

Sr no	Date	Closing price		Daily deviations (returns)
1	5/1/2019	55.32		
2	5/2/2019	51.94	-2.74%	0.07%
3	5/3/2019	52.53	0.49%	0.00%
4	5/6/2019	52.69	0.13%	0.00%
5	5/7/2019	52.19	−0.42%	0.00%
6	5/8/2019	51.73	−0.39%	0.00%
7	5/9/2019	52.17	0.37%	0.00%
8	5/10/2019	52.31	0.11%	0.00%
9	5/13/2019	50.22	−1.77%	0.03%
10	5/14/2019	51.73	1.28%	0.02%
11	5/15/2019	52.19	0.39%	0.00%
12	5/16/2019	52.22	0.02%	0.00%
13	5/17/2019	50.07	−1.82%	0.03%
14	5/20/2019	49.36	−0.62%	0.00%
15	5/21/2019	49.90	0.47%	0.00%
16	5/22/2019	49.49	−0.36%	0.00%
17	5/23/2019	48.01	−1.32%	0.02%
18	5/24/2019	48.49	0.43%	0.00%
19	5/28/2019	47.86	−0.57%	0.00%
20	5/29/2019	48.01	0.13%	0.00%
Realized variance in 20 days				0.20%
Realized Volatility in 20 days				4.42%
Number of trading days in a year				252
Realized variance (annualized)				49.15%
Realized volatility (annualized)				70.11%

D. Bag, *Valuation and Volatility*,
https://doi.org/10.1007/978-981-16-1135-3

Table A2 Buy trade daily 2018

Day date	Trade time	Trade price	Trade quantity	Buy symbol
1-Apr-18	9:41:00	203.95	3	ADANIENT
1-Apr-18	9:41:00	203.95	8	ADANIENT
1-Apr-18	9:41:00	203.95	4	ADANIENT
1-Apr-18	9:41:00	203.95	3	ADANIENT
1-Apr-18	9:41:00	203.95	10	ADANIENT
1-Apr-18	9:41:00	203.95	22	ADANIENT
1-Apr-18	9:41:00	203.95	10	ADANIENT
1-Apr-18	9:41:00	84	10	APOLLOTYRE
1-Apr-18	9:41:00	715	1	BATAINDIA
1-Apr-18	9:41:00	715	9	BATAINDIA
1-Apr-18	9:41:00	715	1	BATAINDIA
1-Apr-18	9:41:00	88.4	20	BOMDYEING
1-Apr-18	9:41:00	88.4	40	BOMDYEING
1-Apr-18	9:41:00	353	90	ACCELYA
1-Apr-18	9:07:42	18.05	1000	ARSHIYA
1-Apr-18	9:07:42	18.05	100	ARSHIYA
1-Apr-18	9:07:42	18.05	200	ARSHIYA
1-Apr-18	9:07:42	119	1	CARBORUNIV
1-Apr-18	9:07:42	119	1	CARBORUNIV
1-Apr-18	9:07:42	95.25	8	CROMPGREAV
1-Apr-18	9:07:42	95.25	17	CROMPGREAV
1-Apr-18	9:07:42	51	1	ESCORTS
1-Apr-18	9:07:42	66.7	149	CENTRALBK
1-Apr-18	9:07:42	532.05	50	GODREJPROP
1-Apr-18	9:07:42	532.05	50	GODREJPROP
1-Apr-18	9:41:00	115.25	50	IRB
1-Apr-18	9:41:00	115.25	50	IRB
1-Apr-18	9:41:00	115.25	50	IRB
1-Apr-18	9:41:00	115.25	1	IRB
1-Apr-18	9:41:00	115.25	100	IRB
1-Apr-18	9:41:00	115.25	49	IRB
1-Apr-18	9:41:00	20.55	100	LANCOIN
1-Apr-18	9:41:00	120	10	RAMCOSYS
1-Apr-18	9:41:00	20.9	500	NOIDATOLL
1-Apr-18	9:41:00	20.9	100	NOIDATOLL
1-Apr-18	9:41:00	20.9	50	NOIDATOLL
2-Apr-18	9:07:44	205	5	ADANIENT
2-Apr-18	9:07:44	205	21	ADANIENT
2-Apr-18	9:07:44	205	2	ADANIENT
2-Apr-18	9:07:44	205	20	ADANIENT
2-Apr-18	9:07:44	205	5	ADANIENT

(continued)

Table A2 (continued)

Day date	Trade time	Trade price	Trade quantity	Buy symbol
2-Apr-18	9:07:44	205	10	ADANIENT
2-Apr-18	9:07:44	205	1	ADANIENT
2-Apr-18	9:07:44	205	2	ADANIENT
2-Apr-18	9:07:44	205	1	ADANIENT
2-Apr-18	9:07:44	205	98	ADANIENT
2-Apr-18	9:07:44	205	2	ADANIENT
2-Apr-18	9:07:44	205	48	ADANIENT
2-Apr-18	9:07:44	205	52	ADANIENT
2-Apr-18	9:07:44	205	50	ADANIENT
2-Apr-18	9:07:44	205	25	ADANIENT
2-Apr-18	9:07:44	205	98	ADANIENT
2-Apr-18	9:07:44	205	2	ADANIENT
2-Apr-18	9:07:44	205	5	ADANIENT
2-Apr-18	9:07:44	205	200	ADANIENT
2-Apr-18	9:07:44	205	1	ADANIENT
2-Apr-18	9:07:44	205	15	ADANIENT
2-Apr-18	9:07:44	205	5	ADANIENT
2-Apr-18	9:07:44	85.5	1	APOLLOTYRE
2-Apr-18	9:07:44	718.85	2	BATAINDIA
2-Apr-18	9:07:44	718.85	2	BATAINDIA
2-Apr-18	9:07:44	718.85	2	BATAINDIA
2-Apr-18	9:07:44	718.85	10	BATAINDIA
2-Apr-18	9:07:44	718.85	1	BATAINDIA
2-Apr-18	9:07:44	718.85	8	BATAINDIA
2-Apr-18	9:07:44	718.85	42	BATAINDIA
2-Apr-18	9:07:44	91	1	BOMDYEING
2-Apr-18	9:07:44	91	2	BOMDYEING
2-Apr-18	9:07:44	91	3	BOMDYEING
2-Apr-18	9:07:44	91	4	BOMDYEING
2-Apr-18	9:07:44	91	5	BOMDYEING
2-Apr-18	9:07:44	91	5	BOMDYEING
2-Apr-18	9:07:44	91	10	BOMDYEING
2-Apr-18	9:07:44	362	1	ACCELYA
2-Apr-18	9:07:44	17.95	50	ARSHIYA
2-Apr-18	9:07:44	17.95	50	ARSHIYA
2-Apr-18	9:07:44	117	5	CARBORUNIV
2-Apr-18	9:07:44	93.1	50	CROMPGREAV
2-Apr-18	9:07:44	93.1	194	CROMPGREAV
2-Apr-18	9:07:44	93.1	125	CROMPGREAV
2-Apr-18	9:07:44	93.1	12	CROMPGREAV
2-Apr-18	9:07:44	93.1	14	CROMPGREAV

(continued)

Table A2 (continued)

Day date	Trade time	Trade price	Trade quantity	Buy symbol
2-Apr-18	9:07:44	93.1	130	CROMPGREAV
2-Apr-18	9:07:44	93.1	19	CROMPGREAV
2-Apr-18	9:07:44	93.1	12	CROMPGREAV
2-Apr-18	9:07:44	93.1	11	CROMPGREAV
2-Apr-18	9:07:44	53.1	6	ESCORTS
2-Apr-18	9:07:44	53.1	27	ESCORTS
2-Apr-18	9:07:44	53.1	52	ESCORTS
2-Apr-18	9:07:44	53.1	1	ESCORTS
2-Apr-18	9:07:44	1379.9	5	CMC
2-Apr-18	9:07:44	64.15	13	COREEDUTEC
2-Apr-18	9:07:44	64.15	41	COREEDUTEC
2-Apr-18	9:07:44	64.15	46	COREEDUTEC
2-Apr-18	9:07:44	64.15	26	COREEDUTEC
2-Apr-18	9:07:44	64.15	200	COREEDUTEC
2-Apr-18	9:07:44	64.15	31	COREEDUTEC
2-Apr-18	9:07:44	64.15	200	COREEDUTEC
2-Apr-18	9:07:44	64.15	17	COREEDUTEC
2-Apr-18	9:07:44	64.15	200	COREEDUTEC
2-Apr-18	9:07:44	64.15	885	COREEDUTEC
2-Apr-18	9:07:44	64.15	200	COREEDUTEC
2-Apr-18	9:07:44	64.15	3	COREEDUTEC
2-Apr-18	9:07:44	68	25	CENTRALBK
2-Apr-18	9:07:44	68	19	CENTRALBK
2-Apr-18	9:07:44	68	44	CENTRALBK
2-Apr-18	9:07:44	68	12	CENTRALBK
2-Apr-18	9:07:44	68	2	CENTRALBK
2-Apr-18	9:07:44	68	82	CENTRALBK
2-Apr-18	9:07:44	68	18	CENTRALBK
2-Apr-18	9:07:44	240.1	2	GUJRATGAS
2-Apr-18	9:07:45	175	1	GMDCLTD
2-Apr-18	9:07:45	175	24	GMDCLTD
2-Apr-18	9:07:45	175	2	GMDCLTD
2-Apr-18	9:07:45	175	42	GMDCLTD
2-Apr-18	9:07:45	175	56	GMDCLTD
2-Apr-18	9:07:45	175	45	GMDCLTD
2-Apr-18	9:07:45	570	1	GODREJPROP
2-Apr-18	9:07:44	84	1	INDIACEM
2-Apr-18	9:07:44	84	3	INDIACEM
2-Apr-18	9:07:44	84	22	INDIACEM
2-Apr-18	9:07:44	84	1	INDIACEM
2-Apr-18	9:07:44	84	32	INDIACEM

(continued)

Table A2 (continued)

Day date	Trade time	Trade price	Trade quantity	Buy symbol
2-Apr-18	9:07:44	84	73	INDIACEM
2-Apr-18	9:07:44	84	29	INDIACEM
2-Apr-18	9:07:44	84	30	INDIACEM
2-Apr-18	9:07:44	84	15	INDIACEM
2-Apr-18	9:07:44	66.15	89	IOB
2-Apr-18	9:07:44	66.15	20	IOB
2-Apr-18	9:07:44	66.15	113	IOB
2-Apr-18	9:07:44	66.15	25	IOB
2-Apr-18	9:07:44	66.15	53	IOB
2-Apr-18	9:07:44	66.15	3	IOB
2-Apr-18	9:07:44	66.15	100	IOB
2-Apr-18	9:07:44	66.15	10	IOB
2-Apr-18	9:07:44	66.15	81	IOB
2-Apr-18	9:07:44	66.15	60	IOB
2-Apr-18	9:07:44	66.15	24	IOB
2-Apr-18	9:07:44	66.15	300	IOB
2-Apr-18	9:07:44	66.15	2	IOB
2-Apr-18	9:07:44	175.1	40	INDIANB
2-Apr-18	9:07:44	175.1	10	INDIANB
2-Apr-18	9:07:44	117.4	25	IRB
2-Apr-18	9:07:44	117.4	6	IRB
2-Apr-18	9:07:44	117.4	8	IRB
2-Apr-18	9:07:44	117.4	21	IRB
2-Apr-18	9:07:44	117.4	11	IRB
2-Apr-18	9:07:44	117.4	39	IRB
2-Apr-18	9:07:44	117.4	25	IRB
2-Apr-18	9:07:44	117.4	5	IRB
2-Apr-18	9:07:44	117.4	50	IRB
2-Apr-18	9:07:44	117.4	41	IRB
2-Apr-18	9:07:44	228	3	MOIL
2-Apr-18	9:07:44	607.1	1	PEL
2-Apr-18	9:07:44	607.1	25	PEL
2-Apr-18	9:07:44	6.8	15	ORIENTPPR
2-Apr-18	9:07:44	2552.5	1	OFSS
2-Apr-18	9:07:44	21.6	150	NOIDATOLL
2-Apr-18	9:07:44	292.7	50	TATASPONGE
2-Apr-18	9:07:45	675	5	TORNTPHARM
2-Apr-18	9:07:45	675	50	TORNTPHARM
2-Apr-18	9:07:45	675	5	TORNTPHARM
2-Apr-18	9:07:45	675	3	TORNTPHARM
2-Apr-18	9:07:45	19.5	500	SJVN

Table A3 Sell trade daily 2018

Day date	Trade time	Trade price	Trade quantity	Symbol
1-Apr-18	9:41:00	203.95	3	ADANIENT
1-Apr-18	9:41:00	203.95	8	ADANIENT
1-Apr-18	9:41:00	203.95	4	ADANIENT
1-Apr-18	9:41:00	203.95	3	ADANIENT
1-Apr-18	9:41:00	203.95	10	ADANIENT
1-Apr-18	9:41:00	203.95	22	ADANIENT
1-Apr-18	9:41:00	203.95	10	APOLLOTYRE
1-Apr-18	9:41:00	84.00	10	BATAINDIA
1-Apr-18	9:41:00	715.00	1	BATAINDIA
1-Apr-18	9:41:00	715.00	9	BATAINDIA
1-Apr-18	9:41:00	715.00	1	BOMDYEING
1-Apr-18	9:41:00	88.40	20	BOMDYEING
1-Apr-18	9:41:00	88.40	40	ACCELYA
1-Apr-18	9:41:00	353.00	90	ARSHIYA
1-Apr-18	9:49:42	18.05	1000	ARSHIYA
1-Apr-18	9:49:42	18.05	100	ARSHIYA
1-Apr-18	9:49:42	18.05	200	CARBORUNIV
1-Apr-18	9:49:42	119.00	1	CARBORUNIV
1-Apr-18	9:49:42	119.00	1	CROMPGREAV
1-Apr-18	9:49:42	95.25	8	CROMPGREAV
1-Apr-18	9:49:42	95.25	17	ESCORTS
1-Apr-18	9:49:42	51.00	1	CENTRALBK
1-Apr-18	9:49:42	66.70	149	GODREJPROP
1-Apr-18	9:49:42	532.05	50	GODREJPROP
1-Apr-18	9:49:42	532.05	50	IRB
1-Apr-18	9:41:00	115.25	50	IRB
1-Apr-18	9:41:00	115.25	50	IRB
1-Apr-18	9:41:00	115.25	50	IRB
1-Apr-18	9:41:00	115.25	1	IRB
1-Apr-18	9:41:00	115.25	100	IRB
1-Apr-18	9:41:00	115.25	49	LANCOIN
1-Apr-18	9:41:00	20.55	100	RAMCOSYS
1-Apr-18	9:41:00	120.00	10	NOIDATOLL
1-Apr-18	9:41:00	20.90	500	NOIDATOLL
1-Apr-18	9:41:00	20.90	100	NOIDATOLL
1-Apr-18	9:41:00	20.90	50	ADANIENT
2-Apr-18	9:49:44	205.00	5	ADANIENT
2-Apr-18	9:49:44	205.00	21	ADANIENT
2-Apr-18	9:49:44	205.00	2	ADANIENT
2-Apr-18	9:49:44	205.00	20	ADANIENT

(continued)

Table A3 (continued)

Day date	Trade time	Trade price	Trade quantity	Symbol
2-Apr-18	9:49:44	205.00	5	ADANIENT
2-Apr-18	9:49:44	205.00	10	ADANIENT
2-Apr-18	9:49:44	205.00	1	ADANIENT
2-Apr-18	9:49:44	205.00	2	ADANIENT
2-Apr-18	9:49:44	205.00	1	ADANIENT
2-Apr-18	9:49:44	205.00	98	ADANIENT
2-Apr-18	9:49:44	205.00	2	ADANIENT
2-Apr-18	9:49:44	205.00	48	ADANIENT
2-Apr-18	9:49:44	205.00	52	ADANIENT
2-Apr-18	9:49:44	205.00	50	ADANIENT
2-Apr-18	9:49:44	205.00	25	ADANIENT
2-Apr-18	9:49:44	205.00	98	ADANIENT
2-Apr-18	9:49:44	205.00	2	ADANIENT
2-Apr-18	9:49:44	205.00	5	ADANIENT
2-Apr-18	9:49:44	205.00	200	ADANIENT
2-Apr-18	9:49:44	205.00	1	ADANIENT
2-Apr-18	9:49:44	205.00	15	ADANIENT
2-Apr-18	9:49:44	205.00	5	APOLLOTYRE
2-Apr-18	9:49:44	85.50	1	BATAINDIA
2-Apr-18	9:49:44	718.85	2	BATAINDIA
2-Apr-18	9:49:44	718.85	2	BATAINDIA
2-Apr-18	9:49:44	718.85	2	BATAINDIA
2-Apr-18	9:49:44	718.85	10	BATAINDIA
2-Apr-18	9:49:44	718.85	1	BATAINDIA
2-Apr-18	9:49:44	718.85	8	BATAINDIA
2-Apr-18	9:49:44	718.85	42	BOMDYEING
2-Apr-18	9:49:44	91.00	1	BOMDYEING
2-Apr-18	9:49:44	91.00	2	BOMDYEING
2-Apr-18	9:49:44	91.00	3	BOMDYEING
2-Apr-18	9:49:44	91.00	4	BOMDYEING
2-Apr-18	9:49:44	91.00	5	BOMDYEING
2-Apr-18	9:49:44	91.00	5	BOMDYEING
2-Apr-18	9:49:44	91.00	10	ACCELYA
2-Apr-18	9:49:44	362.00	1	ARSHIYA
2-Apr-18	9:49:44	17.95	50	ARSHIYA
2-Apr-18	9:49:44	17.95	50	CARBORUNIV
2-Apr-18	9:49:44	117.00	5	CROMPGREAV
2-Apr-18	9:49:44	93.10	50	CROMPGREAV
2-Apr-18	9:49:44	93.10	194	CROMPGREAV
2-Apr-18	9:49:44	93.10	125	CROMPGREAV
2-Apr-18	9:49:44	93.10	12	CROMPGREAV

(continued)

Table A3 (continued)

Day date	Trade time	Trade price	Trade quantity	Symbol
2-Apr-18	9:49:44	93.10	14	CROMPGREAV
2-Apr-18	9:49:44	93.10	130	CROMPGREAV
2-Apr-18	9:49:44	93.10	19	CROMPGREAV
2-Apr-18	9:49:44	93.10	12	CROMPGREAV
2-Apr-18	9:49:44	93.10	11	ESCORTS
2-Apr-18	9:49:44	53.10	6	ESCORTS
2-Apr-18	9:49:44	53.10	27	ESCORTS
2-Apr-18	9:49:44	53.10	52	ESCORTS
2-Apr-18	9:49:44	53.10	1	CMC
2-Apr-18	9:49:44	1379.90	5	COREEDUTEC
2-Apr-18	9:49:44	64.15	13	COREEDUTEC
2-Apr-18	9:49:44	64.15	41	COREEDUTEC
2-Apr-18	9:49:44	64.15	46	COREEDUTEC
2-Apr-18	9:49:44	64.15	26	COREEDUTEC
2-Apr-18	9:49:44	64.15	200	COREEDUTEC
2-Apr-18	9:49:44	64.15	31	COREEDUTEC
2-Apr-18	9:49:44	64.15	200	COREEDUTEC
2-Apr-18	9:49:44	64.15	17	COREEDUTEC
2-Apr-18	9:49:44	64.15	200	COREEDUTEC
2-Apr-18	9:49:44	64.15	885	COREEDUTEC
2-Apr-18	9:49:44	64.15	200	COREEDUTEC
2-Apr-18	9:49:44	64.15	3	CENTRALBK
2-Apr-18	9:49:44	68.00	25	CENTRALBK
2-Apr-18	9:49:44	68.00	19	CENTRALBK
2-Apr-18	9:49:44	68.00	44	CENTRALBK
2-Apr-18	9:49:44	68.00	12	CENTRALBK
2-Apr-18	9:49:44	68.00	2	CENTRALBK
2-Apr-18	9:49:44	68.00	82	CENTRALBK
2-Apr-18	9:49:44	68.00	18	GUJRATGAS
2-Apr-18	9:49:44	240.10	2	GMDCLTD
2-Apr-18	9:49:45	175.00	1	GMDCLTD
2-Apr-18	9:49:45	175.00	24	GMDCLTD
2-Apr-18	9:49:45	175.00	2	GMDCLTD
2-Apr-18	9:49:45	175.00	42	GMDCLTD
2-Apr-18	9:49:45	175.00	56	GMDCLTD
2-Apr-18	9:49:45	175.00	45	GODREJPROP
2-Apr-18	9:49:45	570.00	1	INDIACEM
2-Apr-18	9:49:44	84.00	1	INDIACEM
2-Apr-18	9:49:44	84.00	3	INDIACEM
2-Apr-18	9:49:44	84.00	22	INDIACEM
2-Apr-18	9:49:44	84.00	1	INDIACEM

(continued)

Table A3 (continued)

Day date	Trade time	Trade price	Trade quantity	Symbol
2-Apr-18	9:49:44	84.00	32	INDIACEM
2-Apr-18	9:49:44	84.00	73	INDIACEM
2-Apr-18	9:49:44	84.00	29	INDIACEM
2-Apr-18	9:49:44	84.00	30	INDIACEM
2-Apr-18	9:49:44	84.00	15	IOB
2-Apr-18	9:49:44	66.15	89	IOB
2-Apr-18	9:49:44	66.15	20	IOB
2-Apr-18	9:49:44	66.15	113	IOB
2-Apr-18	9:49:44	66.15	25	IOB
2-Apr-18	9:49:44	66.15	53	IOB
2-Apr-18	9:49:44	66.15	3	IOB
2-Apr-18	9:49:44	66.15	100	IOB
2-Apr-18	9:49:44	66.15	10	IOB
2-Apr-18	9:49:44	66.15	81	IOB
2-Apr-18	9:49:44	66.15	60	IOB
2-Apr-18	9:49:44	66.15	24	IOB
2-Apr-18	9:49:44	66.15	300	IOB
2-Apr-18	9:49:44	66.15	2	INDIANB
2-Apr-18	9:49:44	175.10	40	INDIANB
2-Apr-18	9:49:44	175.10	10	IRB
2-Apr-18	9:49:44	117.40	25	IRB
2-Apr-18	9:49:44	117.40	6	IRB
2-Apr-18	9:49:44	117.40	8	IRB
2-Apr-18	9:49:44	117.40	21	IRB
2-Apr-18	9:49:44	117.40	11	IRB
2-Apr-18	9:49:44	117.40	39	IRB
2-Apr-18	9:49:44	117.40	25	IRB
2-Apr-18	9:49:44	117.40	5	IRB
2-Apr-18	9:49:44	117.40	50	IRB
2-Apr-18	9:49:44	117.40	41	MOIL
2-Apr-18	9:49:44	228.00	3	PEL
2-Apr-18	9:49:44	607.10	1	PEL
2-Apr-18	9:49:44	607.10	25	ORIENTPPR
2-Apr-18	9:49:44	6.80	15	OFSS
2-Apr-18	9:49:44	2552.50	1	NOIDATOLL
2-Apr-18	9:49:44	21.60	150	TATASPONGE
2-Apr-18	9:49:44	292.70	50	TORNTPHARM
2-Apr-18	9:49:45	675.00	5	TORNTPHARM
2-Apr-18	9:49:45	675.00	50	TORNTPHARM
2-Apr-18	9:49:45	675.00	5	TORNTPHARM
2-Apr-18	9:49:45	675.00	3	SJVN
2-Apr-18	9:49:45	19.50	500	ADANIENT

Table A4 Sell trade daily 2019

Day date	Trade time	Trade price	Trade quantity	Symbol
30-Sep-19	12:41:12 AM	76.55	9	IRB
30-Sep-19	12:41:12 AM	76.55	1	IRB
30-Sep-19	12:41:12 AM	76.55	81	IRB
30-Sep-19	12:41:12 AM	76.55	38	IRB
30-Sep-19	12:41:12 AM	76.55	14	IRB
30-Sep-19	12:41:12 AM	76.55	200	IRB
30-Sep-19	12:41:12 AM	76.55	12	IRB
30-Sep-19	12:41:12 AM	76.55	10	IRB
30-Sep-19	12:41:12 AM	76.55	15	IRB
30-Sep-19	12:41:12 AM	76.55	25	IRB
30-Sep-19	12:41:12 AM	76.55	1	IRB
30-Sep-19	12:41:12 AM	76.55	5	IRB
30-Sep-19	12:41:12 AM	76.55	10	IRB
30-Sep-19	12:41:12 AM	76.55	7	IRB
30-Sep-19	12:41:13 AM	595	5	MONSANTO
30-Sep-19	12:41:13 AM	595	5	MONSANTO
30-Sep-19	12:41:13 AM	8.3	600	MARKSANS
30-Sep-19	12:41:13 AM	8.3	1000	MARKSANS
30-Sep-19	12:41:13 AM	8.3	1000	MARKSANS
30-Sep-19	12:41:13 AM	8.3	900	MARKSANS
30-Sep-19	12:41:13 AM	234.5	10	MANDHANA
30-Sep-19	12:41:13 AM	198.7	18	MOIL
30-Sep-19	12:41:13 AM	198.7	5	MOIL
30-Sep-19	12:41:13 AM	198.7	15	MOIL
30-Sep-19	12:41:13 AM	198.7	24	MOIL
30-Sep-19	12:41:13 AM	198.7	2	MOIL
30-Sep-19	12:41:13 AM	198.7	40	MOIL
30-Sep-19	12:41:12 AM	8.15	100	ORIENTPPR
30-Sep-19	12:41:12 AM	66.6	50	PLETHICO
30-Sep-19	12:41:12 AM	122	13	PRESTIGE
30-Sep-19	12:41:12 AM	122	1	PRESTIGE
30-Sep-19	12:41:12 AM	122	4	PRESTIGE
30-Sep-19	12:41:12 AM	122	2	PRESTIGE
30-Sep-19	12:41:12 AM	122	20	PRESTIGE
30-Sep-19	12:41:12 AM	122	71	PRESTIGE
30-Sep-19	12:41:12 AM	122	1	PRESTIGE
30-Sep-19	12:41:12 AM	122	27	PRESTIGE
30-Sep-19	12:41:12 AM	265	1	TATASPONGE
30-Sep-19	12:41:12 AM	265	10	TATASPONGE
30-Sep-19	12:41:12 AM	265	50	TATASPONGE

(continued)

Table A4 (continued)

Day date	Trade time	Trade price	Trade quantity	Symbol
30-Sep-19	12:41:12 AM	265	1	TATASPONGE
30-Sep-19	12:41:12 AM	437	11	TORNTPHARM
30-Sep-19	12:41:12 AM	437	11	TORNTPHARM
30-Sep-19	12:41:12 AM	437	191	TORNTPHARM
30-Sep-19	12:41:12 AM	437	8	TORNTPHARM
30-Sep-19	12:41:12 AM	162.7	18	UNICHEMLAB
30-Sep-19	12:41:12 AM	162.7	18	UNICHEMLAB
30-Sep-19	12:41:12 AM	162.7	25	UNICHEMLAB
30-Sep-19	12:41:12 AM	162.7	31	UNICHEMLAB
30-Sep-19	12:41:12 AM	42.75	1	SMARTLINK
30-Sep-19	12:41:12 AM	34	2	SUPRAJIT
30-Sep-19	12:41:12 AM	54.6	1	SHALPAINTS
30-Sep-19	12:41:12 AM	54.6	100	SHALPAINTS
30-Sep-19	12:41:12 AM	54.6	2	SHALPAINTS
30-Sep-19	12:41:12 AM	54.6	1	SHALPAINTS
30-Sep-19	12:41:12 AM	54.6	1	SHALPAINTS
30-Sep-19	12:41:12 AM	54.6	1	SHALPAINTS
30-Sep-19	12:41:12 AM	19.15	500	SJVN
30-Sep-19	12:41:12 AM	19.15	500	SJVN
30-Sep-19	12:41:12 AM	56	145	TI
30-Sep-19	12:41:12 AM	56	5	TI
30-Sep-19	12:41:12 AM	56	2	TI
30-Sep-19	12:41:12 AM	56	136	TI
30-Sep-19	12:41:12 AM	56	5	TI
30-Sep-19	12:41:12 AM	56	124	TI
30-Sep-19	12:41:12 AM	56	2	TI
30-Sep-19	12:41:12 AM	56	2	TI
30-Sep-19	12:41:12 AM	56	1	TI
30-Sep-19	12:41:12 AM	56	119	TI
30-Sep-19	12:41:12 AM	56	5	TI
30-Sep-19	12:41:12 AM	56	111	TI
30-Sep-19	12:41:12 AM	56	2	TI
30-Sep-19	12:41:12 AM	56	100	TI
30-Sep-19	12:41:12 AM	56	5	TI
30-Sep-19	12:41:12 AM	131	100	TBZ

Table A5 Buy trade daily 2019

Day date	Trade time	Trade price	Trade quantity	Symbol
30-Sep-17	15:30:12 AM	76.55	9	IRB
30-Sep-17	15:30:12 AM	76.55	1	IRB
30-Sep-17	15:30:12 AM	76.55	81	IRB
30-Sep-17	15:30:12 AM	76.55	38	IRB
30-Sep-17	15:30:12 AM	76.55	14	IRB
30-Sep-17	15:30:12 AM	76.55	200	IRB
30-Sep-17	15:30:12 AM	76.55	12	IRB
30-Sep-17	15:30:12 AM	76.55	10	IRB
30-Sep-17	15:30:12 AM	76.55	15	IRB
30-Sep-17	15:30:12 AM	76.55	25	IRB
30-Sep-17	15:30:12 AM	76.55	1	IRB
30-Sep-17	15:30:12 AM	76.55	5	IRB
30-Sep-17	15:30:12 AM	76.55	10	IRB
30-Sep-17	15:30:12 AM	76.55	7	IRB
30-Sep-17	15:30:13 AM	595	5	MONSANTO
30-Sep-17	15:30:13 AM	595	5	MONSANTO
30-Sep-17	15:30:13 AM	8.3	600	MARKSANS
30-Sep-17	15:30:13 AM	8.3	1000	MARKSANS
30-Sep-17	15:30:13 AM	8.3	1000	MARKSANS
30-Sep-17	15:30:13 AM	8.3	900	MARKSANS
30-Sep-17	15:30:13 AM	234.5	10	MANDHANA
30-Sep-17	15:30:13 AM	198.7	18	MOIL
30-Sep-17	15:30:13 AM	198.7	5	MOIL
30-Sep-17	15:30:13 AM	198.7	15	MOIL
30-Sep-17	15:30:13 AM	198.7	24	MOIL
30-Sep-17	15:30:13 AM	198.7	2	MOIL
30-Sep-17	15:30:13 AM	198.7	40	MOIL
30-Sep-17	15:30:12 AM	8.15	100	ORIENTPPR
30-Sep-17	15:30:12 AM	66.6	50	PLETHICO
30-Sep-17	15:30:12 AM	122	13	PRESTIGE
30-Sep-17	15:30:12 AM	122	1	PRESTIGE
30-Sep-17	15:30:12 AM	122	4	PRESTIGE
30-Sep-17	15:30:12 AM	122	2	PRESTIGE
30-Sep-17	15:30:12 AM	122	20	PRESTIGE
30-Sep-17	15:30:12 AM	122	71	PRESTIGE
30-Sep-17	15:30:12 AM	122	1	PRESTIGE
30-Sep-17	15:30:12 AM	122	27	PRESTIGE
30-Sep-17	15:30:12 AM	265	1	TATASPONGE
30-Sep-17	15:30:12 AM	265	10	TATASPONGE
30-Sep-17	15:30:12 AM	265	50	TATASPONGE

(continued)

Table A5 (continued)

Day date	Trade time	Trade price	Trade quantity	Symbol
30-Sep-17	15:30:12 AM	265	1	TATASPONGE
30-Sep-17	15:30:12 AM	437	11	TORNTPHARM
30-Sep-17	15:30:12 AM	437	11	TORNTPHARM
30-Sep-17	15:30:12 AM	437	191	TORNTPHARM
30-Sep-17	15:30:12 AM	437	8	TORNTPHARM
30-Sep-17	15:30:12 AM	162.7	18	UNICHEMLAB
30-Sep-17	15:30:12 AM	162.7	18	UNICHEMLAB
30-Sep-17	15:30:12 AM	162.7	25	UNICHEMLAB
30-Sep-17	15:30:12 AM	162.7	31	UNICHEMLAB
30-Sep-17	15:30:12 AM	42.75	1	SMARTLINK
30-Sep-17	15:30:12 AM	34	2	SUPRAJIT
30-Sep-17	15:30:12 AM	54.6	1	SHALPAINTS
30-Sep-17	15:30:12 AM	54.6	100	SHALPAINTS
30-Sep-17	15:30:12 AM	54.6	2	SHALPAINTS
30-Sep-17	15:30:12 AM	54.6	1	SHALPAINTS
30-Sep-17	15:30:12 AM	54.6	1	SHALPAINTS
30-Sep-17	15:30:12 AM	54.6	1	SHALPAINTS
30-Sep-17	15:30:12 AM	19.15	500	SJVN
30-Sep-17	15:30:12 AM	19.15	500	SJVN
30-Sep-17	15:30:12 AM	56	145	TI
30-Sep-17	15:30:12 AM	56	5	TI
30-Sep-17	15:30:12 AM	56	2	TI
30-Sep-17	15:30:12 AM	56	136	TI
30-Sep-17	15:30:12 AM	56	5	TI
30-Sep-17	15:30:12 AM	56	124	TI
30-Sep-17	15:30:12 AM	56	2	TI
30-Sep-17	15:30:12 AM	56	2	TI
30-Sep-17	15:30:12 AM	56	1	TI
30-Sep-17	15:30:12 AM	56	119	TI
30-Sep-17	15:30:12 AM	56	5	TI
30-Sep-17	15:30:12 AM	56	111	TI
30-Sep-17	15:30:12 AM	56	2	TI
30-Sep-17	15:30:12 AM	56	100	TI
30-Sep-17	15:30:12 AM	56	5	TI
30-Sep-17	15:30:12 AM	131	100	TBZ

Lightning Source UK Ltd.
Milton Keynes UK
UKHW020826270223
417728UK00007B/687